shop mathematics

Dennis C. Ebersole

Prentice-Hall, Inc., Englewood Cliffs, N.J. 07632

Library of Congress Cataloging in Publication Data

EBERSOLE, DENNIS.
    Shop mathematics.

    Includes index.
    1. Shop mathematics. I. Title.
TJ1165.E28   512'.1   79-9275
ISBN 0-13-809038-6

Printed in the United States of America

10 9 8 7 6 5 4 3 2 1

Editorial production supervision
    and interior design by Ian List
Cover design by Frederick Charles, Ltd.
Manufacturing buyer: Gordon Osbourne

Prentice-Hall International, Inc., *London*
Prentice-Hall of Australia Pty. Limited, *Sydney*
Prentice-Hall of Canada, Ltd., *Toronto*
Prentice-Hall of India Private Limited, *New Delhi*
Prentice-Hall of Japan, Inc., *Tokyo*
Prentice-Hall of Southeast Asia Pte. Ltd., *Singapore*
Whitehall Books Limited, *Wellington, New Zealand*

table of contents

preface

This text is an outgrowth of the author's four years experience teaching a course called *Applied Math* at Northampton County Area Community College. The chapters have all been tested in this course and students achieved at least 80% mastery on each chapter.

The text contains many examples illustrated step by step to help the student comprehend each concept. After each section there is an exercise set to test the student's understanding of that section and to provide practice in the use of the concepts. In addition, at the end of each chapter there is a self-test that tests all the concepts taught in the chapter, so the student knows if the concepts have been mastered before taking the chapter test. With over 800 exercises the student receives ample practice in "doing" mathematics. Since the answers to the exercises appear in the back of the text, the student receives immediate feedback on the work.

The text also includes numerous applied problems that require the student to use the mathematical concepts of the chapter in problems from various fields (e.g., problems for machinists, automotive technicians, carpenters, electricians). The instructor can choose those problems most important to the students. Hints or examples are provided for many of the problems to aid the student. Answers to these problems appear in the instructor's manual.

The basic premise behind the instruction in each chapter is that the student should be given a fairly broad mathematical background, even though some of the concepts may have no immediate use. The author's feeling is that the greater knowledge will be utilized in subtle ways if the student has truly comprehended the concepts taught. The instruction is pragmatic, however, and not encumbered with unnecessary mathematical terminology or structure (e.g., set theory is de-emphasized and the axioms of geometry are not treated as a separate entity but are introduced as needed to build up to geometric concepts necessary in the shop).

The important procedures and concepts of each section are "blocked" so that the student will find the text ideal for the purpose of review. In addition, the diagnostic pretests at the beginning of each chapter can be used for a quick review of the chapter as well as a guide of where to look to learn a certain concept. In fact, the pretests serve a threefold purpose. Besides facilitating chapter reviews, they outline the objectives of the unit for the student and can be used by the student to make individual assignments. If the problem on the pretest from a specific section can be done, the instruction in that section may be skipped; otherwise the section should be worked through.

The text is ideal for self-study since each topic is treated fully in a step-by-step manner utilizing programmed instruction. The individual learning the material on his own will find the immediate feedback to each of his responses—usually a complete solution to the problem—as well as the numerous examples and clear exposition on the procedures to use in solving the various types of problems are all conducive to self-study.

The text is also excellent for use in classes. Most classes in shop math are not homogeneous; they are comprised of students with variable rates of learning. The usual textbooks are designed for use with the traditional lecture-discussion technique in which the instructor teaches to the "average" student, and students who are either too far above or below the mean lose interest in the course. In this text I have attempted to alleviate this problem by the use of programmed instruction. This gives the instructor several options in conducting the class.

The instructor can set up an individualized, self-paced course in which the students work through the material at their own rate with mastery obtained on each chapter before going on to the next chapter. If the class is designed to handle students in several areas (e.g., machine shop and automotive technology), the instructor can use a combination modular-contract approach where each student is required to finish certain basic core modules or chapters and can choose from the remaining chapters those which best suit his or her needs. Under such a system the number of units mastered by the student would determine the grade. Finally, an approach that is ideal for teaching shop mathematics is to design shop projects that correlate with the mathematics modules so a group of students learn the mathematical concepts in order to understand and complete the shop project. For example, a group of five to seven students could be given an engine that had to be re-bored because of cylinder wear. As part of the project they would be required to determine the original displacement of the engine and the new displacement after the cylinders had been rebored. In addition, they could be asked

to determine the horsepower at 2200 rpms from the formula $H.P. = \dfrac{PLAN}{396,000}$,

and the S.A.E. horsepower both before and after the reboring. In order to understand and complete this project they would have to learn to use an inside micrometer (Chapter 5) and need the concepts taught in the chapters on applied geometry, decimals, and algebra (formulas). Those students who already knew how to use the micrometer could demonstrate their mastery on a test without working through the chapter.

Northampton County Area Community College   Dennis C. Ebersole
Bethlehem, Pennsylvania

introduction

In any study of mathematics there must be a starting point, a foundation upon which you build. This foundation is usually a set of axioms, a set of mathematical statements that are assumed to be true without proof. The following axioms will be assumed throughout this text.

1. *The commutative law for addition*—this law states that the order in which two numbers are added does not affect their sum; e.g., $6 + 3$ and $3 + 6$ have the same answer.

2. *The commutative law for multiplication*—this law states that the order in which two numbers are multiplied does not affect their product; e.g., 6×3 and 3×6 have the same answer.

3. *The associative law for addition*—this law states that the way in which three numbers being added are grouped does not affect their sum; e.g., $(6 + 3) + 2$ and $6 + (3 + 2)$ have the same answer.

4. *The associative law for multiplication*—this law states that the way in which three numbers being multiplied are grouped does not affect their product; e.g., $(6 \times 3) \times 2$ and $6 \times (3 \times 2)$ have the same answer.

5. *The distributive law*—this law states that the product of one number and the sum of two other numbers is the same as the sum of the products of the first number and each of the addends; e.g., $3(7 + 4)$ and $(3 \times 7) + (3 \times 4)$ have the same answer.

6. *The multiplication (division) law for fractions*—this law states that the fraction obtained by multiplying (dividing) both the numerator and denominator of a fraction by the same number is equal to the original fractions; e.g., (a) (multiplication) $\dfrac{3}{7} = \dfrac{3 \times 5}{7 \times 5} = \dfrac{15}{35}$

$$(b)\ (division) \qquad \frac{15}{21} = \frac{15 \div 3}{21 \div 3} = \frac{5}{7}$$

7. *The addition law for equations*—this law states that the same number can be added to (subtracted from) both sides of an equation without changing the solution to the equation. (See Chapter 6 for examples.)

8. *The multiplication law for equations*—this law states that both sides of an equation can be multiplied (divided) by the same number without changing the solution to the equation.

This text also assumes that you know how to perform the four basic operations with respect to the whole numbers. Since you may know much more than this minimum prerequisite, the following procedure is recommended to ensure that you do not spend too much time on unneeded material, while ensuring that you receive enough instruction in areas that you have not mastered.

1. *Pretest*:   Each chapter begins with a pretest keyed to sections in the chapter. The pretest will diagnose your strengths and weaknesses in that chapter. Next to each problem you will find frame numbers that tell you where in the chapter the concept being tested is taught. Thus, in Pretest 1, the first concept is taught in frames 1 through 6. The answers to the pretest can be found grouped by chapter at the back of the book. First, do all the problems you can; then check your answers. If you do not miss any problems, go to the chapter self-test (see below). If you miss any problems, go to the first frame indicated next to the first problem missed and start the instruction.

2. *Instruction*:   When you find an area in which you need more instruction, work through the instruction. After a brief explanation of a concept you will be asked a question related to the concept. Write your answer; then check to the right of the instruction. The correct answer is given to you. (It is suggested that you cover the answers until ready to check your answer.) If you are correct, proceed to the next frame. If not, try to determine why you were wrong. Continue in this way through the section to the exercise set.

3. *Exercise Set*:   These problems test the concepts taught in the section. Again the answers are supplied, this time in the back of the book. After completing the exercise set, go to the next section in which you need additional work. When you have gone through every necessary section, go to the self-test.

4. *Self-Test*:   The chapter self-test has problems to test all the concepts covered in the chapter. If you can do all of these problems, or correct any mistakes you make, you are ready either to go on to the next chapter or to try the applied problems. If you cannot find your error on any problem, review the appropriate section of the chapter before going on. The answers to the self-test are at the back of the book.

5. *Applied Problems*:   These are problems that require the use of concepts taught in the chapter. They require that you understand the concepts fully and apply them to a problem in a field such as automotive technology, machine technology, or building construction. The answers to these problems are not given to you but are contained in an instructor's manual.

6.   Now go to the next chapter. Begin with the pretest and go through the same cycle until you have mastered all the chapters.

1

fractions

OBJECTIVES

After completing this chapter the student should be able to

1. Define the terms *numerator, denominator, improper fraction, proper fraction, prime, composite.*
2. Write an improper fraction as a mixed number in lowest terms.
3. Write a mixed number as an improper fraction.
4. Find the prime factorization of any counting number.
5. Reduce any fraction to lowest terms.
6. Change a given fraction to an equivalent fraction with a specified denominator.
7. Find the product of a series of fractions.
8. Find the reciprocal of a given numeral.
9. Find the quotient of two fractions.
10. Find the least common multiple of several counting numbers.
11. Find the sum of several fractions or mixed numbers with like or unlike denominators.
12. Find the difference of two fractions or mixed numbers with like or unlike denominators.

If you miss any of these problems, work through the frames indicated next to the problems missed.

(1–6)   **1.** (a) In $\frac{7}{3}$, _____ represents the number of parts into which the whole has been divided.

  (b) Is $\frac{19}{21}$ a proper or an improper fraction?

(7–14)   **2.** (a) Change $7\frac{3}{11}$ to an improper fraction.

  (b) Change $\frac{78}{9}$ to a mixed number.

(15–29)   **3.** (a) Find all the whole number factors of 12.

  (b) Is 27 prime?

  (c) Find the prime factorization of 162.

(30–38)   **4.** (a) Reduce $\frac{75}{105}$ to lowest terms.

  (b) Change $\frac{5}{27}$ to an equal fraction with a denominator of 162.

(39–47)   **5.** Find $2\frac{4}{9} \times \frac{15}{16} \times \frac{24}{25}$.

(48–53)   **6.** Find the reciprocal of each of the following.

  (a) $\frac{13}{4}$

  (b) 5

(54–60)   **7.** Find $1\frac{7}{9} \div 2\frac{2}{5}$.

(61–67)   **8.** Find the least common multiple (L.C.M.) of 12, 30, and 135.

(68–71)   **9.** (a) Find $\frac{13}{18} + \frac{1}{15}$.

  (b) Find $\frac{4}{7} - \frac{3}{28}$.

(72–86)   **10.** (a) Add $4\frac{7}{9}$
            $+ \ 2\frac{1}{3}$

  (b) Subtract 6
            $- \ 3\frac{7}{8}$

1-1   DEFINITIONS

If a 5-inch piece is sawed off the end of a board 8 feet long, how long a piece is left? To find out, you must solve the subtraction problem 8 feet − 5 inches. The answer is not 3 feet or 3 feet-inches since we cannot subtract the 5 inches until both units are the same. We must consider the units in doing the problems. In working with fractions, such as $\frac{3}{8}$, the bottom number (8 in $\frac{3}{8}$) indicates the unit or name of the fraction, while the top number indicates how many of those units you are considering. Remember this and you will have a better understanding of techniques used in dealing with fractions.

1. Fractions are used to express parts of a whole. For example,
    $\frac{1}{2}$ represents 1 part of 2 equal parts that make up a whole, as
    illustrated in Fig. 1-1.

FIGURE 1-1

In the same way $\frac{2}{3}$ represents 2 parts of 3 equal parts that make up
a whole; $\frac{5}{7}$ represents ___5___ parts of ___7___ equal parts that   5; 7
make up a whole.

2. In the fraction $\frac{3}{5}$, the top number, 3, is called the *numerator* and the
    bottom number, 5, is called the *denominator*.
    ___10___ is the denominator in $\frac{7}{10}$.   10

3. ___8___ is the numerator in $\frac{8}{5}$.   8

4. The denominator (or "namer") indicates the number of equal parts
    that make up a whole, while the numerator (or "numberer") indicates
    how many parts you are considering. Which number is the
    numerator in the fraction $\frac{13}{12}$? 13   13

5. Notice that if the numerator is smaller than the denominator, the
    fraction does not make up a whole; while if it is equal to or larger
    than the denominator, the fraction is equal to or greater than a whole
    or 1. For example, 3 thirds or $\frac{3}{3}$ is equal to a whole or 1, while 4
    thirds ($\frac{4}{3}$) is greater than 1. A **proper fraction** is one whose value is
    less than 1. An **improper fraction** is one whose value is greater than
    or equal to 1.

    Examples:
    (a) $\frac{4}{4}$ is improper.
    (b) $\frac{3}{11}$ is proper.
    (c) $\frac{13}{6}$ is improper.

    State whether each of the following are proper or improper fractions.
    (a) $\frac{19}{21}$ (a) proper
    (b) $\frac{16}{3}$ (b) improper
    (c) $\frac{7}{7}$ (c) improper
    (d) $\frac{4}{5}$ (d) proper

6. A mixed number is a number such as $2\frac{3}{5}$ which has both a whole
    number part (2 in $2\frac{3}{5}$) and a fractional part ($\frac{3}{5}$ in $2\frac{3}{5}$). Which of $\frac{7}{3}$, $4\frac{1}{7}$,
    and 6 is a mixed number? $4\frac{1}{7}$

Exercise Set 1-1. Work all the problems before checking your answers.

1. $\frac{5}{6}$ represents _____5_____ parts of _____6_____ equal parts that make up a whole.

_____ ; _____

2. In $\frac{7}{5}$, _____7_____ is the numerator and _____5_____ is the denominator.

_____ ; _____

3. State whether each of the following are proper or improper fractions.
    (a) $\frac{7}{11}$ P
    (b) $\frac{3}{5}$ P
    (c) $\frac{6}{6}$ I
    (d) $\frac{19}{8}$ I

_____
_____
_____
_____

4. Which of $\frac{3}{4}$, 7, $\frac{9}{5}$ and $1\frac{7}{8}$ is a mixed number?

_____

1-2 CHANGING IMPROPER FRACTIONS TO MIXED NUMBERS AND MIXED NUMBERS TO IMPROPER FRACTIONS

To add, subtract, multiply and divide mixed numbers it is often necessary to change a mixed number to a fraction or an improper fraction to a mixed number. You may be able to understand these conversions more readily by considering a steel rule. In the steel rule below notice that there are thirty-two $\frac{1}{32}''$ in $1''$. Thus, $1'' = \frac{32}{32}''$. Similarly, there are sixty-four $\frac{1}{32}''$ in $2''$, so $2'' = \frac{64}{32}''$. Notice also that $1\frac{3}{32}'' = \frac{32}{32}'' + \frac{3}{32}'' = \frac{35}{32}''$.

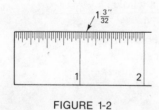

FIGURE 1-2

7. From the introduction above you know that $1\frac{3}{32} = \frac{32}{32} + \frac{3}{32} = \frac{35}{32}$. Similarly, $1\frac{5}{16} = \left(\frac{16}{16}\right) + \frac{5}{16} = \frac{21}{16}$ and $2\frac{3}{8} = \left(\frac{16}{8}\right) + \frac{3}{8} = \frac{19}{8}$. Notice that if the denominator of the fraction is multiplied by the whole number in each mixed number, you will obtain the numerator of the circled fraction above. For example, $16 \times 1 = 16$, and $8 \times 2 = 16$. Notice also that the second fraction is just the fractional part of the mixed number. Thus, $1\frac{3}{16}'' = \frac{16 \times 1}{16} + \frac{3}{16}$ and $2\frac{5}{8} = \frac{8 \times 2}{8} + \frac{5}{8}$.

$7\frac{3}{4} = \frac{4 \times 7}{4} + \frac{3}{4}$. Write $3\frac{5}{7}$ in this form. $3\frac{5}{7} = \frac{21}{7} + \frac{5}{7}$ $3\frac{5}{7} = \frac{7 \times 3}{7} + \frac{5}{7}$

8. $3\frac{5}{7} = \frac{7 \times 3}{7} + \frac{5}{7}$

$= \frac{(7 \times 3) + 5}{7}$

$= \frac{21 + 5}{7} = $ _____

$\frac{26}{7}$

9. $5\frac{6}{11} = \frac{11 \times 5}{11} + \frac{6}{11}$

$= $ _____

$\frac{(11 \times 5) + 6}{11} = \frac{61}{11}$

10.

> To change a mixed number to an improper fraction, multiply the whole number by the denominator of the fraction, add this product to the numerator of the fraction, and put the sum over the denominator.

Example: $7\frac{3}{5} = \frac{(5 \times 7) + 3}{5} = \frac{38}{5}$

Change $5\frac{3}{10}$ to an improper fraction.

$\frac{(10 \times 5) + 3}{10} = \frac{53}{10}$

11. Change $6\frac{3}{7}$ to an improper fraction.

$\frac{7 \times 6 + 3}{7} = \frac{45}{7}$

12. On the steel rule in figure 1.2 $\frac{10}{8} = \frac{8}{8} + \frac{2}{8} = 1 + \frac{2}{8}$ and $\frac{35}{16} = \frac{32}{16} + \frac{3}{16} = 2 + \frac{3}{16}$. Notice that $10 \div 8 = 1$ with a remainder of 2, while $35 \div 16 = 2$ with a remainder of 3. This fact may be used to derive a technique for changing an improper fraction to a mixed number:

> *To change an improper fraction to a mixed number*, divide the numerator by the denominator. The number of times it divides evenly into the numerator is the whole number part of the mixed number. The fractional part is found by placing the remainder over the denominator and reducing to lowest terms.

Example: Change $\frac{21}{16}$ to a mixed number.

$$16\overline{)21} \quad \begin{array}{r} 1\ R5 \\ \underline{16} \\ 5 \end{array}$$

$\frac{21}{16} = 1\frac{5}{16}$

Change $\frac{38}{32}$ to a mixed number.

$\frac{38}{32} = 1\frac{6}{32} = 1\frac{3}{16}$

13. Change $\frac{43}{16}$ to a mixed number.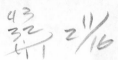

$2\frac{11}{16}$

$$16\overline{\smash{\big)}43} \atop \underline{32} \atop 11$$

with quotient 2 above.

14. Change $\frac{54}{16}$ to a mixed number.

$3\frac{3}{8}$

Exercise Set 1-2.   Work all the problems before checking your answers.

1. Change each mixed number to an improper fraction.
    (a) $5\frac{7}{8}$
    (b) $3\frac{7}{15}$
    (c) $1\frac{13}{17}$

2. Change each improper fraction to a mixed number in lowest terms.
    (a) $\frac{10}{7}$
    (b) $\frac{48}{12}$
    (c) $\frac{78}{14}$

1-3 PRIME FACTORIZATION

One of the most important theorems of arithmetic states that every whole number larger than 1 can be written in just one way as the product of prime numbers. In this section we define prime numbers and show one method of finding this unique prime factorization.

15. The divisors (factors) of a number are the whole numbers that divide evenly into the number. For example, the divisors of 12 are 1, 2, 3, 4, 6, and 12 since these are the only whole numbers that divide evenly into 12. The easiest way to find these factors is start with 1 and then try larger and larger numbers, writing down those numbers that divide evenly into the number and how often they go into the number in pairs until all possibilities are exhausted.

Example:
    (a) 1 divides into 12 twelve times. Write down 1, 12.
    (b) 2 divides into 12 six times. Add 2, 6 to obtain 1, 12, 2, 6.

    (c)   3 divides into 12 four times. Add 3, 4 to find 1, 12, 2, 6, 3, 4.

    (d)   4 divides into 12 three times but we already have both
          factors, so we are done.
          Find the factors of 18.   1, 18, 2, 9, 3, 6

1 18 9 6 3 2

16. Find the factors of 20.   1, 20, 2, 10, 4, 5

20 1 2 2 10 4 5

17. Find the factors of 32.   1, 32, 2, 16, 4, 8

1 32 2 16 4 8

18. Before we look at the prime factorization of a number we must define
a **prime number**. A number is prime if it has exactly two factors or
divisors. For example, 2 and 3 are prime because the only factors are
1 and 2 and 1 and 3, respectively. Is 4 a prime number?   No, it has 3 factors, 1, 2, and 4.

19. Is 5 a prime number?   Yes, since the only factors are 1
and 5.

20. Is 6 a prime number?   No, factors are 1, 2, 3, and 6.

21. Notice that 1 is not a prime number, since it has only one factor, 1.
The first three primes are 2, 3, and 5. Find the next five primes.   *2, 3, 5* 7, 11, 13, 17, 19

22. A number with more than two factors is **composite**. Is 21 composite
or prime?   Composite, factors are 1, 3, 7, 21.

23. We now return to the prime factorization of a number. A
factorization of a number is an indicated multiplication that equals
the number. For example, several factorizations of 18 are

$$1 \times 18 = 18$$
$$2 \times 9 = 18$$
$$2 \times 3 \times 3 = 18$$
$$1 \times 3 \times 6 = 18$$

A **prime factorization** is the indicated product of prime numbers. For
example, the prime factorization of 18 is $2 \times 3 \times 3$ since the product
is 18 and each of the factors is a prime.
Which of the following is the prime factorization of 15?
(a) 1×15 or (b) 3×5 $3 \times 5 = 15$

nonprime #'s expressed as
product of prime #'s
so what?

24. Consider the prime factorization of 15, 3 × 5. Both 3 and 5 divide evenly into 15. Also, if we divide 15 by 3, the quotient is 5 and 3 × 5 = 15. Thus, a way to find a factorization of a number is to divide the number by a prime. Consider the number 20. The prime 2 divides evenly into 20 with a quotient of 10. Also, 2 × 10 is a factorization of 20 but not a prime factorization since 10 is not prime. However, the prime 2 divides evenly into 10 with a quotient of 5. Thus, 10 = 2 × 5 and also 20 = 2 × 10 = 2 × 2 × 5. Notice that 2 × 2 × 5 is the prime factorization of 20.
This is one method for finding the prime factorization of a number.

Example: Find the prime factorization of 75. Since 3 is the first prime that divides evenly into 75, we divide 75 by 3 and write the quotient below 75.

$$3\underline{)75}$$
$$25$$

Then consider 25. 5 is the first prime that divides evenly into 25.

$$3\underline{)75}$$
$$5\underline{)25}$$
$$5$$

The prime factorization is then 3 × 5 × 5. Find the prime factorization of 12.

12 = 2 × 2 × 3
$$2\underline{)12}$$
$$2\underline{)\ 6}$$
$$3$$

25. Finding the prime factorization:
    (a) Try each prime number until you find one that divides evenly into the number.
    (b) Divide the number by this prime and indicate the quotient below the number.
    (c) Continue in this way until the quotient is a prime number.
    (d) The prime factorization is the product of the divisors and the last quotient.

Find the prime factorization of 10.

10 = 2 × 5
$$2\underline{)10}$$
$$5$$

26. Find the prime factorization of 96.

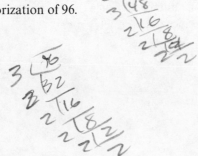

96 = 2 × 2 × 2 × 2 × 2 × 3
$$2\underline{)96}$$
$$2\underline{)48}$$
$$2\underline{)24}$$
$$2\underline{)12}$$
$$2\underline{)\ 6}$$
$$3$$

27. Find the prime factorization of 385.

$385 = 5 \times 7 \times 11$

$5 \underline{|385}$
$7 \underline{|\ 77}$
11

28. The prime factorization of 7 is 7 because 7 is prime. What is the prime factorization of 13?

13

29. Find the prime factorization of 45.

$45 = 3 \times 3 \times 5$

$3\underline{|45}$
$3\underline{|15}$
5

Exercise Set 1-3. Work all the problems before checking your answers.

1. Find the factors of 24.

2. State whether each number is prime or composite.
   (a) 19
   (b) 91
   (c) 21

3. Find the prime factorization of each number.
   (a) 25
   (b) 17
   (c) 125
   (d) 84

1-4 REDUCING FRACTIONS AND CHANGING TO A HIGHER DENOMINATOR

Many fractions have the same value. For example, $\frac{1}{2}$, $\frac{2}{4}$, $\frac{7}{14}$, and $\frac{13}{26}$ are all equal. In order to avoid confusion by having many different looking answers, a convention has been adopted where everyone agrees to write fractional answers reduced to lowest terms. In this section you will learn a method for doing this. In addition, you will learn how to change a fraction to an equal fraction with a larger denominator. This ability is necessary for you to be able to add and subtract fractions with unlike denominators. We use the fact that if we multiply or divide both the numerator and denominator by the same number, the resulting fraction is equal to the original fraction.

---

30. A fraction is said to be reduced to lowest terms if there is no number (except 1) that is a factor of both the numerator and the denominator. For example, $\frac{26}{39}$ is not reduced to lowest terms because 13 is a factor of both 26 and 39. $\frac{10}{15}$ is not reduced to lowest terms because _____ is a factor of both 10 and 15.

5

31. If there is a number other than 1 that is a factor of both the numerator and denominator of a fraction, there must be a prime number that is a factor of both the numerator and denominator since every number has a prime factorization. For example, since 6 is a factor of both the numerator and denominator of $\frac{12}{18}$, both 2 and 3 are factors of 12 and 18 since the prime factorization of 6 is 2×3. Thus, to reduce a fraction to lowest terms we would have to eliminate the prime factors 2 and 3 from both the numerator and denominater. If we divide both the numerator and denominator by the same number, it will not change the value of the fraction. Thus, we could reduce $\frac{12}{18}$ by dividing 12 and 18 by 6.

$$\frac{12 \div 6}{18 \div 6} = \frac{2}{3}$$

Since 4 is a factor of both 20 and 24, reduce $\frac{20}{24}$ by dividing 20 and 24 by 4.

$$\frac{20 \div 4}{24 \div 4} = \frac{5}{6}$$

32. If when you first look at the fraction you do not see what factor is common to both the numerator and denominator, you will need a better method for reducing fractions. This is one place where the prime factorization is used.

> To reduce a fraction:
> (a) Write both numerator and denominator in prime factored form.
> (b) Divide numerator and denominator by the prime factors that appear in both the numerator and denominator.
> (c) Multiply the remaining factors to find the reduced fraction.

Example: Reduce $\frac{36}{40}$.

(a) $\dfrac{36}{40} = \dfrac{2 \times 2 \times 3 \times 3}{2 \times 2 \times 2 \times 5}$

(b) 2 appears twice in the numerator and three times in the denominator so we can eliminate two factors of 2. To do this we divide both numerator and denominator by 2×2.

$$\frac{36}{40} = \frac{(2 \times 2 \times 3 \times 3) \div (2 \times 2)}{(2 \times 2 \times 2 \times 5) \div (2 \times 2)} = \frac{3 \times 3}{2 \times 5}$$

(c) $\dfrac{3 \times 3}{2 \times 5} = \dfrac{9}{10}$

Reduce $\frac{27}{36}$.

(a) $\dfrac{27}{36} = \dfrac{3 \times 3 \times 3}{2 \times 2 \times 3 \times 3}$

(b) $\dfrac{(3 \times 3 \times 3) \div (3 \times 3)}{(2 \times 2 \times 3 \times 3) \div (3 \times 3)}$

$= \dfrac{3}{2 \times 2}$

(c) $\dfrac{3}{2 \times 2} = \dfrac{3}{4}$

33. Reduce $\frac{39}{91}$

(a) $\dfrac{39}{91} = \dfrac{3 \times 13}{7 \times 13}$

(b) $\dfrac{(3 \times 13) \div 13}{(7 \times 13) \div 13} =$

(c) $\dfrac{3}{7}$

34. Reduce $\frac{21}{63}$.

(a) $\dfrac{21}{63} = \dfrac{3 \times 7}{3 \times 3 \times 7}$

(b) $\dfrac{(3 \times 7) \div (3 \times 7)}{(3 \times 3 \times 7) \div (3 \times 7)} =$

(c) $\dfrac{1}{3}$

35. Reduce $\frac{98}{252}$.

(a) $\dfrac{98}{252} = \dfrac{2 \times 7 \times 7}{2 \times 2 \times 3 \times 3 \times 7}$

(b) $\dfrac{(2 \times 7 \times 7) \div (2 \times 7)}{(2 \times 2 \times 3 \times 3 \times 7) \div (2 \times 7)} =$

(c) $\dfrac{7}{18}$

36. Just as we must change pounds to ounces in order to add 3 pounds + 5 ounces, since the units must be the same, to add two fractions we must change them to the same denominator. When we change 3 pounds to ounces, we must multiply the number of pounds by 16 since there are 16 ounces in a pound. In much the same way, in order to change the fraction $\frac{3}{7}$ to a denominator of 28, we must multiply the number of sevenths, 3, by 4, because there are four 7's in 28. To change a fraction to a higher denominator the following procedure is used.

(a) Divide the new denominator by the original denominator.
(b) Multiply numerator and denominator of the fraction by this quotient.

Since we multiply both the numerator and denominator by the same number, the new fraction is equal to the old fraction.

Example: Change $\frac{5}{14}$ to a denominator of 84.

(a) $84 \div 14 = 6$

(b) $\dfrac{5 \times 6}{14 \times 6} = \dfrac{30}{84}$

Change $\frac{7}{15}$ to a denominator of 75.

$$\frac{35}{75}$$

(a) $75 \div 15 = 5$

(b) $\dfrac{7 \times 5}{15 \times 5} = \dfrac{35}{75}$

37. Change $\frac{4}{9}$ to a denominator of 108.

$$\frac{48}{108}$$

(a) $108 \div 9 = 12$

(b) $\dfrac{4 \times 12}{9 \times 12} = \dfrac{48}{108}$

38. Change $\frac{11}{13}$ to a denominator of 156

$$\frac{132}{156}$$

(a) $156 \div 13 = 12$

(b) $\dfrac{11 \times 12}{13 \times 12} = \dfrac{132}{156}$

Exercise Set 1-4.   Work all the problems before checking you answers.

1. Reduce the following fractions to lowest terms.
   (a) $\frac{21}{49}$
   (b) $\frac{15}{90}$
   (c) $\frac{98}{154}$
   (d) $\frac{26}{130}$

   _____
   _____
   _____
   _____

2. Change the following fractions to the indicated denominator.
   (a) $\frac{34}{41}$ to 164
   (b) $\frac{11}{12}$ to 96
   (c) $\frac{10}{27}$ to 405
   (d) $\frac{3}{8}$ to 168

   _____
   _____
   _____
   _____

1-5   MULTIPLICATION OF FRACTIONS AND MIXED NUMBERS

In machining a piece of work from $\frac{3}{8}$ pound of steel, $\frac{1}{9}$ of the metal is scrapped. Do you know the weight of the scrap metal? To determine this you would have to find the product of the two fractions $\frac{3}{8}$ and $\frac{1}{9}$. In this section we shall look at the process used in multiplying fractions.

---

39. The multiplication of two fractions is defined as the product of the numerators over the product of the denominators.

Example: $\frac{3}{7} \times \frac{5}{8} = \frac{3 \times 5}{7 \times 8} = \frac{15}{56}$

Find $\frac{7}{9} \times \frac{4}{11}$. $\frac{28}{99}$

$\frac{7}{9} \times \frac{4}{11} = \frac{7 \times 4}{9 \times 11} = \frac{28}{99}$

40. If several fractions are being multiplied, you place the product of all the numerators over the product of all the denominators.

Example: $\frac{2}{3} \times \frac{2}{7} \times \frac{1}{5} = \frac{2 \times 2 \times 1}{3 \times 7 \times 5} = \frac{4}{105}$

Find $\frac{7}{8} \times \frac{3}{5} \times \frac{1}{2}$. $\frac{21}{80}$

$\frac{7}{8} \times \frac{3}{5} \times \frac{1}{2} = \frac{7 \times 3 \times 1}{8 \times 5 \times 2} = \frac{21}{80}$

41. You can often reduce the amount of work you have to do by writing each numeral in prime factored form and reducing as outlined in Section 1-4.

Example: $\frac{14}{15} \times \frac{4}{49} = \frac{14 \times 4}{15 \times 49} = \frac{2 \times 7 \times 2 \times 2}{3 \times 5 \times 7 \times 7}$

Since 7 is a factor of both the numerator and denominator, we divide the numerator and denominator by 7 to find

$$\frac{(2 \times 7 \times 2 \times 2) \div 7}{(3 \times 5 \times 7 \times 7) \div 7} = \frac{2 \times 2 \times 2}{3 \times 5 \times 7} = \frac{8}{105}$$

If we had done the problem without using prime factorization first, we would have found

$$\frac{14}{15} \times \frac{4}{49} = \frac{14 \times 4}{15 \times 49} = \frac{56}{735}$$

It is not obvious how to reduce $\frac{56}{735}$, although it was easy to find the prime factorization before multiplying.

Find $\frac{7}{15} \times \frac{25}{9}$.

$\frac{7}{15} \times \frac{25}{9} = \frac{(7 \times 5 \times 5) \div 5}{(3 \times 5 \times 3 \times 3) \div 5}$
$= \frac{7 \times 5}{3 \times 3 \times 3} = \frac{35}{27}$

42. Multiply $\frac{24}{35} \times \frac{5}{42}$.

$$\frac{24}{35} \times \frac{5}{42}$$

$$= \frac{(2 \times 2 \times 2 \times 3 \times 5) \div (2 \times 3 \times 5)}{(5 \times 7 \times 2 \times 3 \times 7) \div (2 \times 3 \times 5)}$$

$$= \frac{2 \times 2}{7 \times 7} = \frac{4}{49}$$

43. Find $\frac{7}{18} \times \frac{10}{12} \times \frac{4}{35}$.

$$\frac{7}{18} \times \frac{10}{12} \times \frac{4}{35}$$

$$= \frac{(7 \times 2 \times 5 \times 2 \times 2) \div (2 \times 2 \times 2 \times 5 \times 7)}{(2 \times 3 \times 3 \times 2 \times 2 \times 3 \times 5 \times 7) \div (2 \times 2 \times 2 \times 5 \times 7)}$$

$$= \frac{1}{3 \times 3 \times 3} =$$

$\frac{1}{27}$

44. Since $5\frac{1}{7} = \frac{36}{7}$, $5\frac{1}{7} \times \frac{5}{12} = \frac{36}{7} \times \frac{5}{12} = $ _____
    (Write the answer as a mixed number or proper fraction.)

$2\frac{1}{7}$

$$\frac{36}{7} \times \frac{5}{12}$$

$$= \frac{(2 \times 2 \times 3 \times 3 \times 5) \div (2 \times 2 \times 3)}{(7 \times 2 \times 2 \times 3) \div (2 \times 2 \times 3)}$$

$$= \frac{3 \times 5}{7} = \frac{15}{7} = 2\frac{1}{7}$$

45. Any whole number may be written as an improper fraction by writing a denominator of 1 under the number; e.g., $5 = \frac{5}{1}$, since there are five 1's in 5.
    $6 = \frac{6}{1}$, so
    $6 \times \frac{5}{3} = \frac{6}{1} \times \frac{5}{3} = $ _____

10

46. $5\frac{1}{5} \times \frac{2}{13} \times 3\frac{1}{3} =$
    $\frac{26}{5} \times \frac{2}{13} \times \frac{10}{3} = $ _____

$\frac{8}{3} = 2\frac{2}{3}$

47. $7\frac{1}{3} \times 2\frac{2}{11} \times \frac{1}{4} = $ _____

4

Exercise Set 1-5.   Work all the problems before checking your answers.

1. Find the following products.

    (a) $\frac{3}{11} \times \frac{6}{5}$

    (b) $\frac{9}{14} \times \frac{5}{12} \times \frac{8}{15}$

    (c) $\frac{6}{16} \times \frac{3}{45} \times \frac{20}{7}$

    (d) $5\frac{1}{7} \times 3\frac{1}{4}$

    (e) $4 \times 7\frac{3}{8}$

    (f) $6\frac{2}{3} \times 3\frac{3}{5} \times \frac{2}{9}$

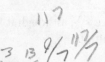

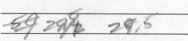

1-6 THE RECIPROCAL

In solving certain types of equations it is necessary to multiply one number by another number that will result in a product of 1. Such numbers are called *reciprocals* of each other. Thus, $\frac{3}{4}$ is the reciprocal of $\frac{4}{3}$ and $\frac{4}{3}$ is the reciprocal of $\frac{3}{4}$, since $\frac{3}{4} \times \frac{4}{3} = 1$. We shall use the concept of a reciprocal in the division of fractions.

---

48. The reciprocal of a number is that number which when multiplied by the first number results in a product of 1. The reciprocal of a fraction can always be found by inverting or "flipping over" the fraction. For example, $\frac{5}{8}$ inverted is $\frac{8}{5}$ and $\frac{5}{8} \times \frac{8}{5} = 1$, so $\frac{8}{5}$ is the reciprocal of $\frac{5}{8}$.
    What is the reciprocal of $\frac{2}{7}$?

    $\frac{7}{2}$

49. What is the reciprocal of $\frac{11}{3}$?

    $\frac{3}{11}$

50. To find the reciprocal of a whole number, we need only note that any whole number can be thought of as that many ones. For example, 7 means 7 ones and 12 means 12 ones. But 7 ones can be written as the fraction $\frac{7}{1}$. Write 12 as a fraction. (Use 1 as the denominator.)

    $\frac{12}{1}$

51. Write 34 as a fraction.

    $\frac{34}{1}$

52. Since 7 can be written as $\frac{7}{1}$, the reciprocal of 7 is found by inverting $\frac{7}{1}$ to obtain $\frac{1}{7}$. What is the reciprocal of 12?

    $\frac{1}{12}$

53. What is the reciprocal of 21?

    $\frac{1}{21}$

Exercise Set 1-6. Work all the problems before checking your answers.

1. Write 9 as a fraction.

2. Find the reciprocal of each of the following.
   (a) 10
   (b) $\frac{1}{4}$
   (c) $\frac{11}{13}$
   (d) $\frac{7}{3}$

1-7 DIVISION OF FRACTIONS AND MIXED NUMBERS

If you wanted to drill five equally spaced holes on a sheet of metal $5\frac{1}{4}$ inches long, what would the center-to-center distance be? To find out you would have to know how to divide mixed numbers.

54. Consider the fraction $\frac{6}{2}$. 6 halves is equal to 3 units. Also, $6 \div 2 = 3$. In fact, the fraction bar, /, is equivalent to the division symbol, $\div$. Thus, 10/3 or $\frac{10}{3}$ can be written as $10 \div 3$. Write $\frac{2}{5}$ using the $\div$ sign.

$2 \div 5$

55. $20 \div 7$ can be written as $\frac{20}{7}$.
Write $5 \div 14$ as a fraction.

$\frac{5}{14}$

56. You are now in position to consider division of fractions. Consider the following problem. $\frac{20}{9} \div \frac{5}{6}$ is equivalent to $\frac{20/9}{5/6}$ from the definition of the fraction bar. We can simplify the denominator, $\frac{5}{6}$, by multiplying by the reciprocal, which will result in a denominator of 1. If we multiply $\frac{5}{6}$ by $\frac{6}{5}$, however, we must also multiply $\frac{20}{9}$ by $\frac{6}{5}$ so we do not change the value of the fraction.

$$\frac{\frac{20}{9} \times \frac{6}{5}}{\frac{5}{6} \times \frac{6}{5}} = \frac{\frac{20}{9} \times \frac{6}{5}}{1}$$

Recall that $7 = \frac{7}{1}$. Therefore,

$$\frac{\frac{20}{9} \times \frac{6}{5}}{1} = \frac{20}{9} \times \frac{6}{5}.$$

We have just shown that $\frac{20}{9} \div \frac{5}{6} = \frac{20}{9} \times \frac{6}{5}$. This leads to the following rule for division.

> To divide fractions or mixed numbers, change the divisor to its reciprocal and multiply.

Example: $\dfrac{15}{4} \div \dfrac{3}{8} = \dfrac{15}{4} \times \dfrac{8}{3}$

$$= \frac{(3 \times 5 \times 2 \times 2 \times 2) \div (2 \times 2 \times 3)}{(2 \times 2 \times 3) \div (2 \times 2 \times 3)} = 10$$

Find $\frac{4}{9} \div \frac{2}{21}$

$\frac{4}{9} \div \frac{2}{21} = \frac{4}{9} \times \frac{21}{2} = \frac{14}{3}$

57. Find $\frac{4}{3} \div 6$. (Recall: $6 = \frac{6}{1}$.)

$\frac{4}{3} \div 6 =$

$\frac{4}{3} \div \frac{6}{1} =$

$\frac{4}{3} \times \frac{1}{6} = \frac{2}{9}$

58. $3\frac{7}{8} \div 1\frac{15}{16} =$

$\frac{31}{8} \div \frac{31}{16} =$ _____

$\frac{31}{8} \times \frac{16}{31} = 2$

59. $5\frac{2}{7} \div 3\frac{3}{14} =$ _____

$\frac{37}{7} \div \frac{45}{14}$

$\frac{37}{7} \times \frac{14}{45} = \frac{74}{45} = 1\frac{29}{45}$

60. $6 \div 2\frac{1}{4} =$

$\frac{6}{1} \div \frac{9}{4} =$ _____

$\frac{6}{1} \times \frac{4}{9} = \frac{8}{3} = 2\frac{2}{3}$

Exercise Set 1-7. Work all the problems before checking your answers.

1. $\frac{20}{7} \div \frac{15}{2}$ _____

2. $11 \div \frac{22}{5}$ _____

3. $\frac{3}{8} \div 6$ _____

4. $\frac{3}{7} \div \frac{14}{9}$ _____

5. $2\frac{1}{8} \div \frac{17}{18}$ _____

6. $3\frac{3}{4} \div 2\frac{1}{5}$ _____

7. $\frac{5\frac{1}{4}}{7}$ _____

1-8 LEAST COMMON MULTIPLE

Just as 5 dimes and 2 quarters does not equal 7 dimes or 7 quarters and 3 feet plus 4 inches does not equal 7 feet or 7 inches, $\frac{3}{14} + \frac{2}{21}$ does not equal $\frac{5}{35}$, $\frac{5}{14}$, or $\frac{5}{21}$. In order to add quantities with unlike units we must change them to the same units. For example, if we change both the dimes and quarters to half dollars, we obtain 5 dimes + 2 quarters = 1 half dollar + 1 half dollar = 2 half dollars. In the same way we can change fractions to the same denominator in order to add them. In order to find the smallest denominator to which both fractions can be changed, you must find the smallest number that is a multiple of both denominators. This number is called the *least common multiple*.

61. A multiple of a number is any number that can be written as the original number times another whole number. For example, some multiples of 12 are

12 = 12 × 1

24 = 12 × 2

36 = 12 × 3

48 = 12 × 4

A number that is a multiple of two or more different numbers is called a *common multiple*. The smallest whole number that is a common multiple of several numbers is called the *least common multiple*. The four smallest multiples of 8 are

8 = 8 × 1

16 = 8 × 2

24 = 8 × 3

32 = 8 × 4

What is the least common multiple of 8 and 12?

24

62. A very efficient method of finding the least common multiple (L.C.M.) uses the prime factorizations of the numbers. (See Section 1-3.) The method is outlined in the next few frames. The first step is to find the prime factorization of each of the numbers.

    Example: Find the least common multiple of 12 and 8.
      (a)   Find the prime factorizations of 12 and 8.

$$2 \underline{|12}$$
$$2\underline{|6}$$
$$3$$
$$12 = 2 \times 2 \times 3$$
$$2\underline{|8}$$
$$2\underline{|4}$$
$$2$$
$$8 = 2 \times 2 \times 2$$

63. Since $12 = 2 \times 2 \times 3$, any multiple of 12 can be written as $2 \times 2 \times 3$ times some other number. For example, $(2 \times 2 \times 3) \times 5$ is a multiple of 12 because it equals 12×5. In other words, any multiple of 12 has $2 \times 2 \times 3$ as part of its prime factorization. Therefore, the least common multiple of 12 and 8 must have $2 \times 2 \times 3$ as part of its prime factorization. Thus, the second step is to write the prime factorization of the first number. The third step is to look at the prime factorization of the second number and determine if any factors must be included with the prime factorization of the first number in order to find a multiple of the second number.

    Example: Find the least common multiple of 12 and 8.
      (a) $12 = 2 \times 2 \times 3$
           $8 = 2 \times 2 \times 2$
      (b)   L.C.M. $= (2 \times 2 \times 3)$
      (c)   L.C.M. $= (2 \times 2 \times 3) \times 2 = 24$
(The extra factor of 2 was included because 8 has 3 factors of 2 in its prime factorization and there were only 2 factors of 2 in the prime factorization of 12. Notice that $(2 \times 2 \times 3) \times 2 = 12 \times 2$ and $2 \times 2 \times 3 \times 2 = 8 \times 3$, so the number is a multiple of both 12 and 8.)

If there are more than two numbers, we continue in the same way until we obtain the prime factorization of a number that includes the prime factorization of each number as part of its prime factorization. Let's go through finding the least common multiple of 20 and 25 step by step.
(a) Find the prime factorizations 20 and 25.
$20 = $ _____

$25 = $ _____

$$2\underline{|20}$$
$$2\underline{|10}$$
$$5$$
$$20 = 2 \times 2 \times 5$$
$$5\underline{|25}$$
$$5$$
$$25 = 5 \times 5$$

64. (b) and (c). Write the prime factorization of 20 and multiply by any factors of 25 that are still needed to find a multiple of 25.

L.C.M. = $(2 \times 2 \times 5) \times 5$
(*Note*: 25 has 2 factors of 5, so another factor of 5 was needed.)

65. (d) Multiply the factors together to find the least common multiple.
L.C.M. = $2 \times 2 \times 5 \times 5 = $ _____

100

66.
> Technique for finding L.C.M.
> (a) Find the prime factorization of each number.
> (b) Write the prime factorization of the first number.
> (c) Multiply this factorization by any factors of the second number that are needed to find a multiple of that number.
> (d) Continue in this way until you find a prime factorization that is a multiple of each number.
> (e) Find the product of this prime factorization.

Example: Find the L.C.M. of 15, 25, and 45.
(a) $15 = 3 \times 5$
$25 = 5 \times 5$
$45 = 3 \times 3 \times 5$
(b) L.C.M. = $(3 \times 5) \times$
(c) L.C.M. = $(3 \times 5) \times 5 \times$
(A 5 is needed because $25 = 5 \times 5$ has two 5s as part of its prime factorization.)
(d) L.C.M. = $[(3 \times 5) \times 5] \times 3$
(45 has two factors of 3 and one factor of 5. Since we already had one of the 3's and the 5, we need only include an additional 3.)
(e) L.C.M. = $3 \times 5 \times 5 \times 3 = 225$

Find the L.C.M. of 21, 28, and 27.

(a) $21 = 3 \times 7$
$28 = 2 \times 2 \times 7$
$27 = 3 \times 3 \times 3$
(b) L.C.M. = $(3 \times 7) \times$
(c) L.C.M. = $(3 \times 7) \times 2 \times 2$
(d) L.C.M. = $[(3 \times 7) \times 2 \times 2]$
$\times 3 \times 3$
(e) $= 756$

67. Find the L.C.M. of 24 and 36.

(a) $24 = 2 \times 2 \times 2 \times 3$
$36 = 2 \times 2 \times 3 \times 3$
(b) L.C.M. = $(2 \times 2 \times 2 \times 3) \times$
(c) L.C.M. = $(2 \times 2 \times 2 \times 3) \times 3$
(d) $= 72$

Exercise Set 1-8.   Work all the problems before checking your answers.
Find the L.C.M. of the following sets of numbers.

1. 16, 20 _____

2. 9, 4 _____

3. 30, 5, 42 _____

4. 40, 30, 75 _____

5. 55, 22, 10 _____

1-9 ADDITION AND SUBTRACTION OF FRACTIONS

Imagine that you would like to bolt a $\frac{3}{16}$-inch piece of metal onto a $\frac{3}{8}$-inch piece.
If a $\frac{1}{32}$-inch washer and a $\frac{1}{4}$-inch nut are used, what is the minimum under the head
length bolt you can use? The answer to your problem requires the ability to add
fractions.

68. Note that in 3 inches + 4 inches = 7 inches and
7 inches − 2 inches = 5 inches the unit or denomination *inches* does
not change. In the same manner
$\frac{3}{7} + \frac{1}{7}$ = 3 sevenths + 1 seventh = 4 sevenths or $\frac{4}{7}$, while
$\frac{8}{9} - \frac{4}{9} = \frac{4}{9}$. In other words, to add (or subtract) fractions with the
same denominator, add (subtract) the numerators and put the sum
(difference) over the common denominator.

Examples:

(a) $\frac{3}{13} + \frac{5}{13} = \frac{3 + 5}{13} = \frac{8}{13}$

(b) $\frac{14}{15} - \frac{1}{15} = \frac{14 - 1}{15} = \frac{13}{15}$

Add $\frac{2}{9} + \frac{5}{9}$. $\frac{7}{9}$

69. To add (subtract) two fractions with unlike denominators we must
first change each fraction to the same denominator. This is where
the least common multiple and changing to a higher denominator are
used. The least common multiple of the denominators is the number
to which we shall change each of the original denominators. This
number is usually called the *least common denominator* (L.C.D.).

Example: Add $\frac{2}{21} + \frac{3}{14}$.
(a)   Find the L.C.D. of 21 and 14.

$$21 = 3 \times 7$$
$$14 = 2 \times 7$$
$$\text{L.C.D.} = (3 \times 7) \times 2 = 42$$

(b) Change each of the fractions to the higher denominator, the L.C.D. (that is, change $\frac{2}{21}$ and $\frac{3}{14}$ to fractions with a denominator of 42).

$$\frac{2}{21} \times \frac{2}{2} = \frac{4}{42}$$
$$\frac{3}{14} \times \frac{3}{3} = \frac{9}{42}$$

(c) Add the numerators of the new fractions and place the sum over the L.C.D.

$$\frac{2}{21} + \frac{3}{14} = \frac{4}{42} + \frac{9}{42} = \frac{13}{42}$$

Use these three steps to add $\frac{7}{15} + \frac{3}{10}$.

(a) $15 = 3 \times 5$
$10 = 2 \times 5$
L.C.D. $= (3 \times 5) \times 2 = 30$

(b) $\frac{7}{15} \times \frac{2}{2} = \frac{14}{30}$
$\frac{3}{10} \times \frac{3}{3} = \frac{9}{30}$

(c) $\frac{14}{30} + \frac{9}{30} = \frac{23}{30}$

70. | Addition (subtraction) of fractions:
(a) Find the L.C.D.
(b) Change each fraction to an equal fraction with the L.C.D. for a denominator.
(c) Add (subtract) the numerators of the new fraction and place the sum over the L.C.D.
(d) Reduce, if possible.

Example: Subtract $\frac{15}{16} - \frac{5}{12}$.
(a) $16 = 2 \times 2 \times 2 \times 2$
$12 = 2 \times 2 \times 3$
L.C.D. $= 2 \times 2 \times 2 \times 2 \times 3$
$\quad\quad = 48$

(b) $\frac{15 \times 3}{16 \times 3} = \frac{45}{48}; \frac{5 \times 4}{12 \times 4} = \frac{20}{48}$

(c) $\frac{45}{48} - \frac{20}{48} = \frac{25}{48}$

Subtract $\frac{3}{7} - \frac{1}{14}$.

(a) $7 = 7$
$14 = 2 \times 7$
L.C.D. $= 7 \times 2 = 14$

(b) $\frac{3 \times 2}{7 \times 2} = \frac{6}{14}$

$\frac{1 \times 1}{14 \times 1} = \frac{1}{14}$

(c) $\frac{6}{14} - \frac{1}{14} = \frac{5}{14}$

71.   Add $\frac{14}{33} + \frac{3}{22}$.

(a) $33 = 3 \times 11$
     $22 = 2 \times 11$
     L.C.D. $= 3 \times 11 \times 2$
     $\qquad = 66$

(b) $\dfrac{14 \times 2}{33 \times 2} = \dfrac{28}{66}$

     $\dfrac{3 \times 3}{22 \times 3} = \dfrac{9}{66}$

(c) $\dfrac{28}{66} + \dfrac{9}{66} = \dfrac{37}{66}$

Exercise Set 1-9.   Work all the problems before checking your answers.

1. Add $\frac{1}{16} + \frac{3}{20}$.

2. Add $\frac{5}{6} + \frac{1}{18}$.

3. Add $\frac{2}{35} + \frac{3}{49}$.

4. Add $\frac{7}{39} + \frac{5}{26}$.

5. $\frac{20}{27} - \frac{5}{27}$

6. $\frac{5}{6} - \frac{2}{9}$

7. $\frac{10}{13} - \frac{5}{7}$

8. $\frac{8}{77} - \frac{1}{98}$

1-10   ADDITION AND SUBTRACTION OF MIXED NUMBERS

If the total weight of a basket of fruit is 3 pounds 7 ounces and the basket weighs 15 ounces, how would you find the weight of the fruit? To solve this problem you would have to subtract 15 ounces from 3 pounds 7 ounces.

$$
\begin{array}{r}
\text{3 pounds   7 ounces} \\
-\ \text{15 ounces} \\
\hline
\end{array}
$$

We cannot subtract 15 from 7 so something must be changed. Since there are 16 ounces in a pound, we can change the problem to 2 pounds + 16 ounces + 7 ounces minus 15 ounces or

$$
\begin{array}{r}
\text{2 pounds 23 ounces} \\
-\ \text{15 ounces} \\
\hline
\text{2 pounds   8 ounces}
\end{array}
$$

The technique used in solving this problem is exactly analogous to what is done in solving the problem $3\frac{7}{16} - \frac{15}{16}$, which you will learn in this section.

72. Since $2\frac{1}{7} = 2 + \frac{1}{7}$ and

$5\frac{5}{7} = 5 + \frac{5}{7}, 2\frac{1}{7} + 5\frac{5}{7} =$

$2 + \frac{1}{7} + 5 + \frac{5}{7} = (2 + 5) + (\frac{1}{7} + \frac{5}{7})* =$

$7 + \frac{6}{7} = \underline{\hspace{2cm}}.$ $\qquad 7\frac{6}{7}$

73.
> To add mixed numbers, add the whole number parts together and the fractional parts together and simplify the resulting mixed number. If the fractional parts have different denominators, they will have to be changed to the least common denominator.

Example:

$$\begin{array}{r} 2\frac{7}{8} \\ + 1\frac{7}{10} \\ \hline \end{array}$$

L.C.D. = 40

$$\begin{array}{r} 2\frac{7}{8} \\ + 1\frac{7}{10} \\ \hline \end{array} \implies \begin{array}{r} 2\frac{35}{40} \\ + 1\frac{28}{40} \\ \hline 3\frac{63}{40} \end{array}†$$

Since $\frac{63}{40} = 1\frac{23}{40}$,

$3\frac{63}{40} = 3 + \frac{63}{40} = 3 + 1\frac{23}{40} = 3 + 1 + \frac{23}{40} = \underline{\hspace{2cm}}$ $\qquad 4\frac{23}{40}$

74. Add

$$\begin{array}{r} 4\frac{5}{12} \\ + 2\frac{3}{4} \\ \hline \end{array}$$

$\qquad 7\frac{1}{6}$

L.C.D. = 12

$$\begin{array}{r} 4\frac{5}{12} \\ + 2\frac{3}{4} \\ \hline \end{array} \implies \begin{array}{r} 4\frac{5}{12} \\ + 2\frac{9}{12} \\ \hline 6\frac{14}{12} \end{array}$$

$6\frac{14}{12} = 6 + 1\frac{2}{12} = 7\frac{2}{12} = 7\frac{1}{6}$

75. Add

$$\begin{array}{r} 3\frac{2}{9} \\ + 1\frac{1}{6} \\ \hline \end{array}$$

$\qquad 4\frac{7}{18}$

L.C.D. = 18

$$\begin{array}{r} 3\frac{2}{9} \\ + 1\frac{1}{6} \\ \hline \end{array} \implies \begin{array}{r} 3\frac{4}{18} \\ + 1\frac{3}{18} \\ \hline 4\frac{7}{18} \end{array}$$

---

* See the commutative law for addition in Introduction, p. 1

† The symbol, "$\implies$", which means "implies", is used to indicate that the problem on the left is equivalent to the problem to its right.

76. Add

$$5\frac{2}{7}$$
$$+\ 3$$

$8\frac{2}{7}$

77. Add

$$7\frac{2}{5}$$
$$+\ 5\frac{7}{15}$$

$12\frac{13}{15}$

78. | To subtract mixed numbers, subtract the fractional parts first. If the fraction being subtracted is larger than the fraction being subtracted from, you will have to borrow from the whole number part of the mixed number being subtracted from, just as you must borrow from the 8 in the subtraction problem

$$81$$
$$-\ 35$$

Example:

$$3\frac{7}{16}$$
$$-\ \ \frac{15}{16}$$

Since $\frac{15}{16}$ is larger than $\frac{7}{16}$, you will have to borrow from the 3.

$$3\frac{7}{16} \implies 2+1\frac{7}{16} \implies 2\frac{23}{16}$$
$$-\ \ \frac{15}{16} \qquad\quad -\ \ \ \ \frac{15}{16} \qquad\quad -\ \ \frac{15}{16}$$

Finish the problem

$$3\frac{7}{16}$$
$$-\ \ \frac{15}{16}$$

$$2\frac{23}{16}$$
$$-\ \ \frac{15}{16}$$
$$\overline{2\frac{8}{16}} = 2\frac{1}{2}$$

79. Subtract

$$7\frac{5}{8}$$
$$-\ 2\frac{7}{8}$$

$4\frac{3}{4}$

$$7\frac{5}{8} \implies 6+1\frac{5}{8} \implies 6\frac{13}{8}$$
$$-\ 2\frac{7}{8} \qquad\quad -\ \ \ \ 2\frac{7}{8} \qquad\quad -\ 2\frac{7}{8}$$
$$\overline{\qquad\qquad\qquad\qquad\qquad\qquad\quad 4\frac{6}{8}}$$

$$4\frac{6}{8} = 4\frac{3}{4}$$

80. | If the denominators of the fractional parts are not the same, they must be changed to the least common denominator before proceeding.

Example:

$$2\tfrac{7}{8}$$
$$-\ 2\tfrac{4}{7}$$

L.C.D. = 56

$$\begin{array}{c}2\tfrac{7}{8} \\ -\ 2\tfrac{4}{7}\end{array} \implies \begin{array}{c}2\tfrac{49}{56} \\ -\ 2\tfrac{32}{56} \\ \hline \tfrac{17}{56}\end{array}$$

Subtract

$$3\tfrac{5}{11}$$
$$-\ 2\tfrac{1}{2}$$

$$\tfrac{21}{22}$$
L.C.D. = 22

$$\begin{array}{c}3\tfrac{5}{11} \\ -\ 2\tfrac{1}{2}\end{array} \implies \begin{array}{c}3\tfrac{10}{22} \\ -\ 2\tfrac{11}{22}\end{array}$$

$$\begin{array}{c}3\tfrac{10}{22} \\ -\ 2\tfrac{11}{22}\end{array} \implies \begin{array}{c}2\tfrac{32}{22} \\ -\ 2\tfrac{11}{22} \\ \hline \tfrac{21}{22}\end{array}$$

81. Subtract

$$8\tfrac{1}{4}$$
$$-\ 3\tfrac{1}{18}$$

$$5\tfrac{7}{36}$$
L.C.D. = 36

$$\begin{array}{c}8\tfrac{1}{4} \\ -\ 3\tfrac{1}{18}\end{array} \implies \begin{array}{c}8\tfrac{9}{36} \\ -\ 3\tfrac{2}{36} \\ \hline 5\tfrac{7}{36}\end{array}$$

82. Subtract

$$5\tfrac{3}{14}$$
$$-\ \tfrac{8}{21}$$

$$4\tfrac{5}{6}$$

83. Consider the problem

$$5$$
$$-\ 3\tfrac{2}{9}$$

The fractional part of the mixed number 5 is 0. Since we are going to try to subtract $\tfrac{2}{9}$ from 0, we would like the denominator to be 9. Recall that any whole number can be written with a denominator of 1. Thus, $0 = \tfrac{0}{1}$. Change $\tfrac{0}{1}$ to an equal fraction with a denominator of 9.

$$\tfrac{0}{1} \times \tfrac{9}{9} = \tfrac{0}{9}$$
Thus, $0 = \tfrac{0}{9}$.

84. Thus,

$$\begin{array}{r} 5 \\ - 3\frac{2}{9} \end{array} \implies \begin{array}{r} 5\frac{0}{9} \\ - 3\frac{2}{9} \end{array}$$

Solve this problem.

$$4 + 1\frac{0}{9} \implies \begin{array}{r} 4\frac{9}{9} \\ - 3\frac{2}{9} \\ \hline 1\frac{7}{9} \end{array}$$

$$\left(Note: 1\frac{0}{9} = \frac{9 \times 1 + 0}{9} = \frac{9}{9}\right)$$

85. Using the information from frames 83 and 84 we see that if either of the mixed numbers in a subtraction problem has 0 for its fractional part, the fractional part should be rewritten as 0 over the denominator of the fraction in the other mixed number. Then proceed as described before.
Subtract

$$\begin{array}{r} 7 \\ - 2\frac{5}{11} \end{array}$$

$$4\frac{6}{11}$$

$$\begin{array}{r} 7 \\ - 2\frac{5}{11} \end{array} \implies \begin{array}{r} 7\frac{0}{11} \\ - 2\frac{5}{11} \end{array} \implies \begin{array}{r} 6\frac{11}{11} \\ - 2\frac{5}{11} \\ \hline 4\frac{6}{11} \end{array}$$

86. Subtract

$$\begin{array}{r} 6\frac{7}{25} \\ - 4 \end{array}$$

$$\begin{array}{r} 6\frac{7}{25} \\ - 4\frac{0}{25} \\ \hline 2\frac{7}{25} \end{array}$$

Exercise Set 1-10.   Work all the problems before checking your answers.

1. Add

(a) $\begin{array}{r} 3\frac{11}{15} \\ + 2\frac{5}{12} \end{array}$

_____

(b) $\begin{array}{r} 7\frac{1}{16} \\ + 2\frac{1}{12} \end{array}$

_____

(c) $\begin{array}{r} 6 \\ + 3\frac{2}{9} \end{array}$

_____

2. Subtract

(a) $\begin{array}{r} 3\frac{1}{15} \\ - 2\frac{5}{12} \end{array}$

(b) $\begin{array}{r} 4\frac{6}{7} \\ - 2 \end{array}$

_____

(c) $\begin{array}{r} 11 \\ - 8\frac{10}{11} \end{array}$

_____

SELF-TEST

Work all the problems before checking your answers.

1. $\frac{7}{8}$ represents _____ parts of _____ equal parts, which make up a whole.

 _____ ; _____

2. In $\frac{3}{16}$, _____ is the denominator and _____ is the numerator.

 _____ ; _____

3. State whether each of the following is proper or improper.
 (a) $\frac{18}{22}$

 _____

 (b) $\frac{5}{5}$

 _____

 (c) $\frac{6}{5}$

 _____

4. Find the factors of 36.

 _____

5. Is 91 prime or composite?

 _____

6. Find the prime factorization of each number.
 (a) 75

 _____

 (b) 147

 _____

 (c) 19

 _____

7. Reduce the following fractions to lowest terms.
 (a) $\frac{26}{39}$

 _____

 (b) $\frac{8}{20}$

 _____

 (c) $\frac{17}{85}$

 _____

8. Change the following fractions to the indicated denominator.
 (a) $\frac{7}{13}$ to 156

 _____

 (b) $\frac{5}{6}$ to 54

 _____

9. Write $3\frac{1}{7}$ as an improper fraction.

 _____

10. Write $\frac{22}{4}$ as a mixed number in lowest terms.

 _____

11. Find the following products.
 (a) $\frac{7}{8} \times \frac{5}{6}$

 _____

 (b) $\frac{16}{75} \times \frac{15}{12} \times \frac{35}{36}$

 _____

 (c) $3\frac{1}{3} \times 2\frac{1}{5}$

 _____

 (d) $7 \times 2\frac{2}{7} \times \frac{3}{8}$

 _____

12. Find the reciprocal of each of the following.
 (a) $\frac{5}{9}$

 _____

 (b) 6

 _____

13. Divide and reduce the answer to lowest terms.
 (a) $\frac{25}{26} \div \frac{10}{13}$

 _____

 (b) $\frac{14}{15} \div \frac{25}{7}$

 _____

 (c) $4\frac{1}{5} \div 6\frac{2}{3}$

 _____

 (d) $6 \div \frac{1}{3}$

 _____

 (e) $\frac{2\frac{1}{4}}{3}$

 _____

14. Find the least common multiple of the following sets of numbers.
   (a) 18, 30   _____
   (b) 143, 22, 26   _____

15. Add and reduce the answer to lowest terms.
   (a) $\frac{1}{7} + \frac{5}{7}$ _____
   (b) $\frac{6}{25} + \frac{7}{45}$ _____
   (c) $\quad 3$
   $\underline{+ \; 2\frac{1}{7}}$ _____
   (d) $\quad 7\frac{2}{3}$
   $\underline{+ \; 3\frac{1}{3}}$ _____
   (e) $\quad 5\frac{3}{8}$
   $\underline{+ \; 2\frac{9}{14}}$ _____

16. Subtract and reduce to lowest terms.
   (a) $\frac{6}{13} - \frac{2}{13}$ _____
   (b) $\frac{6}{35} - \frac{4}{49}$ _____
   (c) $\quad 2$
   $\underline{- \; 1\frac{3}{8}}$ _____
   (d) $\quad 7\frac{1}{2}$
   $\underline{- \; 5}$ _____
   (e) $\quad 5\frac{8}{9}$
   $\underline{- \; 2\frac{5}{9}}$ _____
   (f) $\quad 4\frac{1}{15}$
   $\underline{- \; 3\frac{1}{12}}$ _____

APPLIED PROBLEMS*

In each of the following problems one of the variables, X, Y, Z, etc., associated with the original problem is missing from the table. Use the concepts of this chapter to find the missing variable.

   A. An engine is rated at X horsepower. Under actual working conditions the engine only delivers Y horsepower, which is Z of the rated horsepower.

_____

*The Applied Problems section shows how you can apply some of the concepts introduced in the chapter to solve problems from various technical fields.
While working through this section you should try to find problems from your career area that could be solved using the mathematical concepts of this chapter.

	X	Y	Z
1.	$\frac{3}{4}$		$\frac{7}{8}$
2.	$\frac{3}{2}$	$\frac{7}{8}$	
3.		$\frac{2}{3}$	$\frac{3}{5}$
4.	$\frac{15}{16}$		$\frac{4}{5}$
5.	$\frac{2}{3}$	$\frac{3}{8}$	

Example: 1. Since the actual horsepower is $\frac{7}{8}$ of the rated horsepower of $\frac{3}{4}$, the actual horsepower is $\frac{7}{8} \times \frac{3}{4} = 21/32$.

B. If the pitch (distance between adjacent threads) of a bolt is X inches, a bolt will advance Y inches in Z revolutions.

	X	Y	Z
1.		$\frac{7}{8}''$	14
2.	$\frac{1}{8}''$		16
3.	$\frac{3}{32}''$	$\frac{3}{4}''$	
4.		$\frac{15}{16}''$	8
5.	$\frac{1}{16}''$		12

Example: 1. If the bolt advances $\frac{7}{8}''$ in 14 revolutions, the advance in one revolution is $\frac{7}{8}'' \div 14 = \frac{7}{8}'' \times \frac{1}{14} = \frac{1}{16}''$.

C. The under-the-head length of a bolt used to bolt a bumper to a frame must be at least W inches if X inches is allowed for a washer and nut, the frame is Y inches thick, and the bumper is Z inches thick.

	W	X	Y	Z
1.		$\frac{1}{2}''$	$\frac{7}{16}''$	$\frac{3}{8}''$
2.		$\frac{3}{8}''$	$\frac{5}{16}''$	$\frac{1}{4}''$
3.	$\frac{13}{8}''$		$\frac{5}{8}''$	$\frac{9}{16}''$
4.	$\frac{31}{16}''$	$\frac{1}{2}''$		$\frac{5}{8}''$
5.		$\frac{7}{16}''$	$\frac{3}{4}''$	$\frac{5}{8}''$

Example: 1.
$$W = X + Y + Z$$
$$= \tfrac{1}{2}'' + \tfrac{7}{16}'' + \tfrac{3}{8}''$$
$$= \tfrac{8}{16}'' + \tfrac{7}{16}'' + \tfrac{6}{16}''$$
$$= \tfrac{21}{16}''$$

D. If the inside diameter of a pipe is X inches and the pipe is Y inches thick, the outside diameter is Z inches.

	X	Y	Z
1.	$\frac{3}{4}''$	$\frac{1}{16}''$	
2.	$\frac{7}{8}''$	$\frac{1}{4}''$	
3.		$\frac{3}{16}''$	$\frac{9}{8}''$
4.		$\frac{1}{8}''$	$\frac{15}{16}''$
5.	$\frac{17}{16}''$		$\frac{19}{16}''$

Example: 1. Since the thickness must be counted twice in determining the outside diameter,

$$Z = \tfrac{3}{4}'' + \tfrac{1}{16}'' + \tfrac{1}{16}''$$
$$= \tfrac{12}{16}'' + \tfrac{1}{16}'' + \tfrac{1}{16}''$$
$$= \tfrac{14}{16}''$$
$$= \tfrac{7}{8}''$$

E. The total resistance, W (ohms), of three resistors in series with resistances of X ohms, Y ohms, and Z ohms is found by adding the resistances together. Find the missing values in the table below.

	W	X	Y	Z
1.	$\frac{7}{4}$	$\frac{3}{8}$	$\frac{7}{8}$	
2.		$\frac{1}{4}$	$\frac{3}{16}$	$\frac{7}{16}$
3.	$\frac{17}{8}$		$\frac{1}{2}$	$\frac{3}{4}$
4.		$\frac{1}{2}$	$\frac{3}{8}$	$\frac{3}{16}$
5.	$\frac{15}{16}$	$\frac{1}{4}$	$\frac{1}{8}$	

F. Find the missing dimensions in Fig. 1-3.

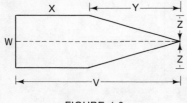

FIGURE 1-3

	V	W	X	Y	Z
1.	$\frac{7}{4}$	$\frac{5}{8}$		$\frac{7}{8}$	
2.			$\frac{3}{8}$	$\frac{7}{16}$	$\frac{1}{4}$
3.	$\frac{13}{8}$	$\frac{3}{4}$		$\frac{11}{16}$	
4.			$\frac{5}{16}$	$\frac{1}{2}$	$\frac{3}{8}$
5.	$\frac{21}{16}$		$\frac{5}{16}$		$\frac{5}{16}$

G. The area, A, of a trapezoid (see Fig. 1-4) can be found by adding the two bases (X and Y), multiplying them by $\frac{1}{2}$, and multiplying this product by the height, Z. Find the missing dimensions in the table.

	A	X	Y	Z
1.	$\frac{9}{64}$	$\frac{5}{8}$	$\frac{1}{2}$	
2.		$\frac{3}{4}$	$\frac{7}{8}$	$\frac{1}{2}$
3.		$\frac{1}{2}$	$\frac{7}{16}$	$\frac{3}{8}$
4.	$\frac{19}{32}$	$\frac{7}{4}$	$\frac{7}{8}$	
5.		$\frac{7}{8}$	$\frac{13}{32}$	$\frac{3}{4}$
6.	$\frac{13}{64}$	$\frac{17}{16}$	$\frac{13}{4}$	

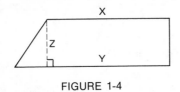

FIGURE 1-4

Example: 1. $\frac{5}{8} + \frac{1}{2} = \frac{5}{8} + \frac{4}{8} = \frac{9}{8}$

$\frac{1}{2} \times \frac{9}{8} = \frac{9}{16}$

$\frac{9}{16} \times Z = \frac{9}{64}$

$Z = \frac{9}{64} \div \frac{9}{16} = \frac{9}{64} \times \frac{16}{9} = \frac{1}{4}$

H.   The sum of the sides X, Y, and Z of a triangle is W.

	W	X	Y	Z
1.		$\frac{3}{4}$	$\frac{7}{16}$	$\frac{3}{8}$
2.		$\frac{5}{9}$	$\frac{7}{15}$	$\frac{12}{25}$
3.	$\frac{21}{8}$	$\frac{5}{6}$	$\frac{3}{4}$	
4.	$\frac{14}{5}$		$\frac{7}{12}$	$\frac{7}{15}$
5.		$\frac{15}{14}$	$\frac{7}{6}$	$\frac{20}{21}$

I.   If Jim finished X pieces in Y hours, he finished an average of Z pieces per hour.

	X	Y	Z
1.	$\frac{7}{4}$	$\frac{1}{2}$	
2.	$\frac{5}{6}$	$\frac{3}{4}$	
3.		$\frac{5}{6}$	$\frac{3}{4}$
4.		$\frac{4}{21}$	$\frac{7}{8}$
5.	$\frac{7}{15}$	$\frac{10}{49}$	
6.	$\frac{11}{12}$	$\frac{7}{18}$	

Example: 1.   The number of pieces per hour, Z, can be found by dividing the number of pieces finished by the number of hours, so $Z = \frac{7}{4} \div \frac{1}{2} = \frac{7}{4} \times \frac{2}{1} = \frac{7}{2}$.

J.   The circumference, C, of a circle is approximately $\frac{22}{7}$ times the diameter, D.

	C	D
1.		$\frac{14}{33}$
2.	$\frac{5}{7}$	
3.		$\frac{21}{22}$
4.	$\frac{16}{3}$	
5.		$\frac{6}{7}$

K. When measuring, the greatest possible error, Y, is always $\frac{1}{2}$ times X, the smallest unit of measurement possible with the tool being used. Thus, the greatest possible error on a ruler marked off in $\frac{1}{2}$ centimeter steps is $\frac{1}{2} \times \frac{1}{2} = \frac{1}{4}$ centimeter. Find the greatest possible error in each case below:

	X	Y
1.	$\frac{1}{16}$ inch	
2.	$\frac{1}{64}$ inch	
3.	$\frac{1}{1,000}$ inch	
4.	$\frac{1}{4}$ centimeter	
5.	10 millimeters	

L. If the tap, Y, for a bolt should be $\frac{3}{4}$ the outside diameter, X, what should the tap be in each case below?

	X	Y
1.	$\frac{3}{4}''$	
2.	$\frac{5}{8}''$	
3.	$1''$	
4.	$\frac{5}{4}''$	
5.	$2''$	

M. If X amps are available for use, Y lamps that pull Z amps each can be used. Find the missing values in the table below.

	X	Y	Z
1.	15		$1\frac{1}{2}$
2.		7	$\frac{3}{4}$
3.	$22\frac{1}{2}$		$2\frac{1}{2}$
4.	$5\frac{1}{7}$	6	
5.	$6\frac{1}{8}$		$\frac{7}{8}$

Example: 1. The problem reduces to finding out how many $1\frac{1}{2}$'s there are in 15. This is a division problem. $15 \div 1\frac{1}{2} = \frac{15}{1} \times \frac{2}{3} = 10$.

N. If a wall is X inches wide by Y inches high, Z rows of bricks are needed to brick the wall from floor to ceiling, and W bricks are needed for each row. (Assume each brick when laid covers $2\frac{1}{2}'' \times 8\frac{3}{4}''$ including the mortar.)

	X	Y	Z	W
1.	$83\frac{1}{8}$	$87\frac{1}{2}$		
2.	$59\frac{1}{16}$	85		
3.	$74\frac{3}{8}$	75		
4.	$91\frac{7}{8}$	$77\frac{1}{2}$		
5.			23	$10\frac{1}{2}$

Example: 1. To find the number of rows, Z, we must divide the height, Y, by the width of one brick, $2\frac{1}{2}''$. Thus, $Z = 87\frac{1}{2} \div 2\frac{1}{2}$, so $Z = \frac{175}{2} \div \frac{5}{2} = \frac{175}{2} \times \frac{2}{5} = 35$. The number of bricks per row, W, can be found by dividing the width of the wall, X, by the length of the brick, $8\frac{3}{4}$. Thus, $W = 83\frac{1}{8} \div 8\frac{3}{4} = \frac{665}{8} \div \frac{35}{4} = \frac{665}{8} \times \frac{4}{35} = 9\frac{1}{2}$.

O. If first X inches and then Y inches are cut off the end of a board $84''$ long, Z inches are left on the board.

	X	Y	Z
1.	$10\frac{1}{2}$	$14\frac{3}{8}$	
2.	$13\frac{1}{8}$	$17\frac{7}{16}$	
3.	$16\frac{1}{16}$	$28\frac{1}{2}$	
4.	$35\frac{7}{8}$	$19\frac{3}{4}$	
5.	$8\frac{3}{16}$		$60\frac{3}{4}$

Example: 1. Since the number of inches left on the board after each cut is less than $84''$ by the amount cut off, Z can be found by subtracting X and Y from $84''$. $84'' - 10\frac{1}{2}'' =$

$$\begin{array}{r} 84 \\ -10\frac{1}{2} \\ \hline \end{array} \implies \begin{array}{r} 83\frac{2}{2} \\ -10\frac{1}{2} \\ \hline 73\frac{1}{2}'' \end{array}$$

$$\begin{array}{r} 73\frac{1}{2} \\ -14\frac{3}{8} \\ \hline \end{array} \implies \begin{array}{r} 73\frac{4}{8} \\ -14\frac{3}{8} \\ \hline 59\frac{1}{8}'' \end{array}$$

Thus, $Z = 59\frac{1}{8}''$.

P. What was the total amount cut off the board in Problem O in each case?

1. _____
2. _____
3. _____
4. _____
5. _____

Q. X rows of bricks are Y meters high, and each brick with mortar is Z meters thick.

	X	Y	Z
1.	12		$\frac{1}{13}$
2.	24	$1\frac{11}{13}$	
3.	16		$\frac{1}{12}$
4.	18	$1\frac{11}{25}$	
5.		$2\frac{1}{2}$	$\frac{1}{14}$
6.		1	$\frac{1}{13}$

Example: 1.   12 rows times $\frac{1}{13}$ meter per row $= \frac{12}{13}$ meter high.

R. There are approximately $2\frac{1}{2}$ centimeters per inch. How many centimeters are there in each case below?

1. 6 inches   _____
2. 1 foot   _____
3. $\frac{1}{2}$ yard   _____
4. $2\frac{1}{2}$ inches   _____
5. $1\frac{3}{5}$ feet   _____

S. Find the missing dimensions for the axle shaft shown in Fig. 1-4A.

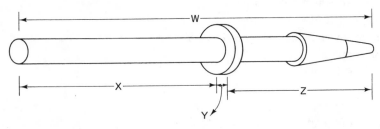

FIGURE 1-4A

	W	X	Y	Z
1.		$15\frac{1}{4}''$	$\frac{3}{8}''$	$8\frac{1}{16}''$
2.	58 cm	$46\frac{1}{2}$ cm	$1\frac{1}{10}$ cm	
3.	$56\frac{1}{6}$ cm		$\frac{9}{10}$ cm	$18\frac{1}{3}$ cm
4.	28''	$16\frac{3}{8}''$		$7\frac{3}{4}''$
5.		$14\frac{5}{8}''$	$\frac{7}{16}''$	$8\frac{1}{16}''$

T. If the shaft makes X revolutions in Y minutes, it is turning at Z revolutions per minute.

	X	Y	Z
1.	3,000	$2\frac{1}{3}$	
2.	$3,130\frac{1}{5}$	3	
3.		$1\frac{1}{10}$	2,100
4.	$637\frac{1}{2}$		1,250
5.		$\frac{5}{12}$	600

U. If a machinist wants to drill 20 holes equally spaced around a disk, or make a gear with 40 teeth, he may use an indexing head. A standard indexing head is set up so that it requires 40 turns of the indexing crank for the material to make one revolution. Thus, if a mark is made after each turn of the crank, 40 equally spaced marks will be made around the material, while if every second turn is marked, 20 equally spaced marks are made. If you wanted 15 equally spaced holes, however, you would have to turn the crank $\frac{40}{15} = 2\frac{2}{3}$ turns before making each mark. To facilitate making a fractional part of a turn such as two-thirds of a turn, plates are used that have 5 or 6 circles formed with different numbers of holes (see Fig. 1-5). Thus, if we are using plate 2 in the table below, in order to make two-thirds of a turn we could use the 39-hole circle in the following way. Since 39 is evenly divisible by 3, $\frac{2}{3} \times 39$ is a whole number, 26. Note that $\frac{26}{39} = \frac{2}{3}$. Thus, if you turned the crank so that the material moved 26 holes in the 39 hole circle, two-thirds of a turn has been made. Thus, by making 2 full turns followed by 26 holes on the 39-hole circle before making a mark, 15 equally spaced marks can be made. This is called the *indexing*. One technique for finding the indexing is illustrated below.

Find the indexing for 25 divisions using one of the plates in the table.
1. $\frac{40}{25} = 1\frac{15}{25} = 1\frac{3}{5}$

2. Both 50 and 60 in plate 3 are divisible by 5, so either could be used. $\frac{3}{5} \times 50 = 30$, while $\frac{3}{5} \times 60 = 36$.

3. The indexing is 1 turn 30 holes in 50-hole circle or 1 turn 36 holes in 60-hole circle.

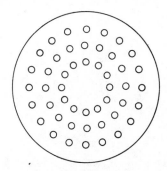

FIGURE 1-5

Plate 1	Plate 2	Plate 3
21 holes	37 holes	48 holes
23 holes	39 holes	50 holes
27 holes	41 holes	51 holes
29 holes	43 holes	54 holes
31 holes	47 holes	56 holes
33 holes	49 holes	60 holes

Find the indexing for the following number of divisions:

1. 32 _____

2. 28 _____

3. 74 _____

4. 22 _____

5. 85 _____

2

decimals

OBJECTIVES

After completing this chapter the student should be able to
1. State the place value of a specified digit in any decimal numeral.
2. Write any decimal numeral in expanded form.
3. Find the sum of decimal numerals.
4. Find the difference of two decimal numerals.
5. Round off a decimal numeral to a specified place value.
6. Determine the number of significant digits in a decimal numeral.
7. Write the answer to a problem involving decimals rounded off to the correct place or correct number of significant digits.
8. Find the product of decimal numerals.
9. Find the quotient of two decimal numerals.
10. Write a fraction as an equivalent decimal numeral.

(1-11)   **1.**   What is the place value of the indicated digits in 81.0765?

    (i) 8   _____

    (ii) 7   _____

    (iii) 5   _____

(12-27)   **2.**   Round off as indicated:

    (a) 72,461 to the nearest thousand.   _____

    (b) .1234 to the nearest hundredth.   _____

    (c) 4.109 to two significant digits.   _____

(28-36)   **3.**   (a) Add $6 + .08 + 3.733$.   _____

    (b) Subtract $25 - .16$.   _____

(37-48)   **4.**   (a) Multiply (_do not round off_) $3.1 \times .036$.   _____

    (b) Multiply and round off to the correct number of digits:

$$.016 \times 3100$$ _____

(49-61)   **5.**   (a) Divide (_do not round off_) $4 \div .025$.   _____

    (b) Divide and round off to the nearest thousandths:

$$1.23 \div 21$$ _____

    (c) Divide and round off to the correct number of significant digits.

$$2.06 \div .0046$$ _____

(62-66)   **6.**   (a) Change $\frac{3}{16}$ to a decimal numeral.   _____

    (b) Change $\frac{17}{13}$ to a decimal numeral rounded off to the nearest hundredth.   _____

2-1 PLACE VALUE

Probably because early man used his fingers to count, we use a number system based on ten digits, the numerals 0, 1, 2, 3, 4, 5, 6, 7, 8, and 9. Since there are only ten digits, to represent any number larger than 9 we must have some convention that uses a combination of these digits as the desired numeral. This convention is based on the concept of place value, the idea that a digit may mean different things if it is in a different _place_. For example, the digit 2 in 213 does not mean the same as the digit 2 in 12. This same concept of place value is used to represent fractions in this decimal notation also.

---

1.   Even though 21 and 12 have the same digits, 1 and 2, in them, they do not represent the same number because of the position of the digits. Our number system is based on ten digits. For this reason the value of a digit increases by a product of 10 as it moves to the left in a numeral. Figure 2-1 indicates these _place values_. We could continue

this chart to include hundred millions, etc., but you will probably have no need for the larger numbers and the pattern continues in the same way.

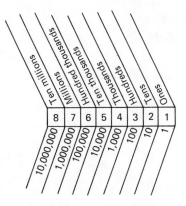

FIGURE 2-1

In the numeral 21, the 2 is in the second or tens place so it means 2 tens, while the 1 is in the ones place so it means 1 one. Thus, in our number system 21 means 2 tens + 1 one.

534 = 5 hundreds + _____ tens + _____ ones.

3; 4

2. Using Fig. 2-1, 4,378 = 4 thousands + 3 hundreds + 7 tens + 8 ones. 62,459 =

6 ten thousands + 2 thousands + 4 hundreds + 5 tens + 9 ones

3. Since the 4 in 4,378 is in the thousands place, its place value is one thousand. What is the place value of the 3?

one hundred

4. What is the place value of 7 in 678,512?

ten thousand

5. Starting at the ones place, the values of the digits increase by powers of ten as you go to the left. To represent a number smaller than one you must go to the right of the ones place. As you go to the right in our number system you must divide by 10 to find the next place value. For example, the place value immediately to the right of the 1,000's place is $1,000 \div 10 = 100$. What is the place value immediately to the right of the 100's place?

$100 \div 10 = 10$

6. Using the technique of frame 5, what is the place value of the place immediately to the right of the ones place?

$1 \div 10 = \frac{1}{10}$

7. What is the place value of the place immediately to the right of the one-tenths place?

$\frac{1}{10} \div 10 = \frac{1}{10} \times \frac{1}{10} = \frac{1}{100}$

8. Since you must have some way of knowing which is the ones place in a decimal numeral in order to be able to say it, a dot, called a *decimal point*, is placed just to the right of the ones place. The decimal point separates the whole number and fractional part of a decimal numeral. Figure 2-2 summarizes what we have discussed so far in this chapter. The "th" is used to indicate a fraction in the terms *tenths, hundredths,* etc.

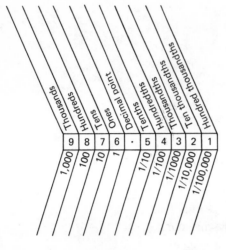

FIGURE 2-2

The place value of the 5 in 9876.54321 is tenths. What is the place value of the 4?   hundredths

9. What is the place value of the 6 in 3.046?   thousandths

10. What is the place value of the 7 in 271.04?   tens

11. What is the place value of the 4 in .00314?   hundred-thousandths

___

Exercise Set 2-1.   Work all the problems before checking your answers.
  1. Write 45,213 in expanded form.
      _____

      _____

  2. What is the place value of the indicated digit in 9,124.83675.
      (a) 9   _____
      (b) 4   _____
      (c) 8   _____
      (d) 6   _____
      (e) 5   _____
      (f) 2   _____
      (g) 7   _____

2-2 ROUNDING OFF DECIMALS AND SIGNIFICANT DIGITS

With the wide use of calculators, students and technicians are often confronted with answers to multiplication and division problems accurate to nine or more places. How valid are these answers? For example, if to the nearest pound 30 bricks of the same weight weighed 148 pounds, if you weighed 10 bricks would you expect their weight to be exactly 49.3333333 pounds? Of course not, since the actual weight of the bricks may have been slightly more or slightly less than 148 pounds. For this reason you must consider how many of the digits in the answer you may expect to be reasonably accurate. These are called the *significant digits*. You must then round off your answer to the correct number of significant digits. In this section you will encounter rules for determining significant digits and for rounding off.

---

12. Most of us round off decimals without really thinking about it. If the sticker price of an automobile is $3,499, you probably refer to it as a $3,500 car. If the odometer reading is 215.8 miles, you probably write down 216 miles. When estimating the total cost of items purchased, a $19.95 pair of slacks is added as $20. Each of these is an example of rounding off. Rounding off is a technique for approximating a decimal numeral to the desired accuracy. One method of rounding off is illustrated below.

    Example: Round off 238 to the nearest ten by the following steps.
    (a) Underline the digit in the tens place and circle the digit immediately to the right of this digit: 23⑧
    (b) If the circled digit is 5 or larger, add 1 to the underlined digit. If it is less than 5, add nothing.

    $$\overset{1}{23\underline{}⑧}$$

    (c) If the circled digit is before the decimal point, replace it and every digit after it that is before the decimal point by a 0. If it is after the decimal point, just drop all the digits after the underlined digit. Then add the 0 or 1 to the remaining digits to obtain your answer.

    $$\overset{1}{23⑧} \implies \overset{1}{230} \implies 240 \text{ (answer)}$$

Is 24 closer to 20 or 30?

20

13. Is 553 closer to 500 or 600?

600, since it is only 47 from 600 but 53 from 500.

14. The reason for the rule that you add 1 to the preceding digit if the circled digit is 5 or larger is illustrated by frames 12 and 13. If the digit is 5 or larger, it is closer to the larger number; while if it is 4 or less, it is closer to the smaller number. Since in rounding off we want to approximate the number, we want the rounded off number to be the closer one. Thus, 2,421 round off to the nearest hundred is _____, not 2,500.

2,400

15. Refer to the steps in frame 12.
Round off 117 to the nearest ten.

(a) 11⑦
            1
(b) 11⑦
(c) 120 (answer)

16. Round off 6048 to the nearest hundred.

(a) 60④8
            0
(b) 60④8
(c) 6,000 (answer)

17. **Example:** Round off .1486 to the nearest hundredth.

(a) .14⑧6
            1
(b) .14⑧6
(c) .15

(*Note*: The 8 and 6 are dropped because they are after the decimal point. This is done to ensure that everyone knows that the answer is only accurate to the nearest hundredth. If the answer is written .1500, where the last digit is in the ten-thousandths place, the answer is assumed to be accurate to the nearest ten-thousandth.)

Round off 1.013 to the nearest hundredth.

(a) 1.01③
            0
(b) 1.01③
(c) 1.01 (answer)

18. In order to state a rule for the number of digits in the best answer the concept of significant digits must be considered. For example, if a newspaper article states that 2,400 people showed up for a concert, did exactly 2,400 people attend? Probably not—the 2,400 is an estimate rounded off to the nearest hundred. Thus, only the first two digits are important or *significant*. Unless you know otherwise, *a*

series of zeros at the end of a number and before the decimal point are not considered to be significant digits. How many significant digits are there in 4,170?

three

19. How many significant digits are there in 400?

one

20. How many significant digits are there in 407?

Three—the zero does not come at the end of the number.

21.

> As a general rule zeros are significant, if
> (a) They are between two nonzero digits, or
> (b) They are at the *end* of a number and at least one of them is *after* the decimal point. In all other cases they are not significant.

Thus, .007 has one significant digit, while .700 has three significant digits. How many significant digits are there in .0031?

Two, because the zeros are at the beginning of the number.

22. How many significant digits are there in 4.006?

Four, because the zeros are between two nonzero digits.

23. How many significant digits are there in .050?

Two, because the second zero is at the end of the number and after the decimal point.

24. In order to express your answer in the best form you must be able to round off to a specific number of significant digits.

Example: Round off 3.1846 to three significant digits.
   (a) Underline the correct significant digit, and circle the digit to the immediate right of it.

$$3.18\underline{4}6$$

   (b) Round off as outlined previously.

$$
\begin{array}{c}
0 \\
3.18\underline{4}6
\end{array}
$$

   3.18 (answer)

Round off .0417 to two significant digits.

(a) .041⑦
(b)   1
    .041⑦
    .042 (answer)

25. Round off .08105 to three significant digits.   .0811

26. Round off 3,698 to three significant digits.   3,700

27. Round off 4.117 to two significant digits.   4.1

Exercise Set 2-2.   Work all the problems before checking your answers.

1. Round off as indicated:
    (a) 2,416 to the nearest hundred   2,400
    (b) 40,489 to the nearest thousand   40,000
    (c) .0309 to the nearest hundredth   .3600
    (d) 319.8 to the nearest one   320
    (e) 6.1234 to the nearest thousandth   6.123
    (f) 3,671 to two significant digits   3,7
    (g) .04681 to three significant digits   .050
    (h) 3.1995 to four significant digits   3.2000

2. How many significant digits are in the following?   2
    (a) 3,400   3
    (b) 20.03   3
    (c) .003

2-3   ADDITION AND SUBTRACTION OF DECIMALS

To understand how to add and subtract decimal numerals you should look at the addition and subtraction of mixed numbers. In adding or subtracting mixed numbers the fractions are added or subtracted separately from the whole numbers, except that you may have to borrow from or carry over a whole number to complete the problem. Also, in the denominators of the fractions, the names or values must be the same before you can add or subtract them. In the same way, decimal numerals must be lined up so that digits with the same place value are lined up vertically when adding or subtracting.

---

28. **Example:**   Add 3.3 + 4.07. Since $3.3 = 3\frac{3}{10}$ and $4.07 = 4\frac{7}{100}$, to add these two numerals we must first change $\frac{3}{10}$ to a denominator of 100 and then add the fractions together and the whole numbers together. Since $\frac{3}{10} = \frac{30}{100}$, we have $3\frac{30}{100} + 4\frac{7}{100} = 7\frac{37}{100}$. To add in decimal form we still change the three-tenths to thirty-hundredths—3.3 is changed to 3.30—and add the fractional parts and then the whole

number parts. To do this we line up the decimal points when adding. Thus, to add 3.3 + 4.07, place the 3.3 above the 4.07 with the decimal points lined up and add a zero at the end of 3.3 to change it to hundredths, also.

$$\begin{array}{r} 3.3 \\ + \; 4.07 \\ \hline \end{array} \implies \begin{array}{r} 3.30 \\ + \; 4.07 \\ \hline 7.37 \end{array}$$

Add 5.6 + 3.14.

$$\begin{array}{r} 5.6 \\ + \; 3.14 \\ \hline \end{array} \implies \begin{array}{r} 5.60 \\ + \; 3.14 \\ \hline 8.74 \end{array}$$

29.

> To add decimal numerals, line up the decimal points and add zeros as needed to ensure that the fractional part of each decimal has the same number of places. Then add from right to left keeping the decimal point in the same position.

Example: Add 5.6771 + .06.

$$\begin{array}{r} 5.6771 \\ + \quad .0600 \\ \hline 5.7371 \end{array}$$

Add 57.09 + 1.0761.

$$\begin{array}{r} 57.0900 \\ + \quad 1.0761 \\ \hline 58.1661 \end{array}$$

30. Add 5 + .36.
(*Note*: Since 5 is a whole number, the decimal point goes after the 5.)

$$\begin{array}{r} 5.00 \\ + \quad .36 \\ \hline 5.36 \end{array}$$

31. **Example:** Find 1.067 + 3 + 29.16.

$$\begin{array}{r} 1.067 \\ 3.000 \\ + \; 29.160 \\ \hline 33.227 \end{array}$$

Add 6.08 + 4.1 + 6.

$$\begin{array}{r} 6.08 \\ 4.10 \\ + \quad 6.00 \\ \hline 16.18 \end{array}$$

32. | Subtraction of decimals also requires that the decimal points be lined up and zeros added, if necessary. |

Example: 5.71 − 4.3

$$\begin{array}{r} 5.71 \\ -\ 4.30 \\ \hline 1.41 \end{array}$$

Subtract   6.1 − .83.

$$\begin{array}{r} \overset{5}{}\overset{1}{0}1 \\ \cancel{6}.\cancel{1}0 \\ -\ .83 \\ \hline 5.27 \end{array}$$

33.   Subtract 5 − 3.461.

$$\begin{array}{r} \overset{9\ 9}{\overset{4\ 1\ 1\ 1}{\cancel{5}.\cancel{0}\cancel{0}0}} \\ -\ 3.461 \\ \hline 1.539 \end{array}$$

34.   Subtract 6.31 − 2.841.

$$\begin{array}{r} 6.310 \\ -\ 2.841 \\ \hline 3.469 \end{array}$$

35.   Consider the problem 3.4 + 1.28 = 4.68. If both the 3.4 and 1.28 are rounded off values, then it is possible that the 3.4 was actually 3.44 and the 1.28 could have been 1.283. In this case the sum would be 4.723 instead of 4.68. Notice that the digits in the tenths place are different in the two answers. When adding or subtracting decimal numerals which have been rounded off (such as measured values) the answer should also be rounded off, since we cannot guarantee the complete accuracy of the answer.

| When adding rounded off decimal numerals the answer should be rounded off in the following way:
 (a)   Determine the place value of the last significant digit in each decimal numeral
 (b)   Round off the answer to the same place as the largest place value determined in step (a). |

Example: In 3.4 + 1.28, the place value of the 4 in 3.4 is tenths, or $\frac{1}{10}$, while the place value of the 8 in 1.28 is hundredths, or $\frac{1}{100}$. Since

$\frac{1}{10}$ is larger than $\frac{1}{100}$, the answer should be rounded off to the nearest tenth.

$$3.4 + 1.28 = 4.68$$

Rounding off, the answer is 4.7.

Find the difference of $3.0 - 1.485$ and round off the answer to the correct place.

$$\begin{array}{r} 3.000 \\ -\ 1.485 \\ \hline 1.515 \end{array}$$

Answer: 1.5 (Rounded Off To The Correct Place)

36. Find the sum of $103.4 + 21$ and round off the answer to the correct place.

124

Exercise Set 2-3.   Work all the problems before checking your answers.

1. Add

    (a) $\begin{array}{r} 4.5 \\ 3.07 \\ +\ 2.105 \\ \hline \end{array}$

    (b) $\begin{array}{r} .6 \\ +\ .9 \\ \hline \end{array}$

    (c) $6 + .81 + 3.1745$

    (d) $.8 + 6.394$ (round off the answer to the correct place)

2. Subtract

    (a) $\begin{array}{r} 4.178 \\ -\ .34 \\ \hline \end{array}$

    (b) $.681 - .1945$ (round off the answer to the correct place)

    (c) $6 - 2.418$

2-4 MULTIPLICATION OF DECIMALS

You can determine the outside circumference or distance around the outside of a gear or wheel by multiplying the outside diameter of the gear or wheel by 3.1416. If the diameter is 6.875″, this will require the multiplication of two decimal numerals. In this section you will learn how to solve this problem and similar problems that require multiplication of decimals.

37. Consider the problem $3.7 \times .003$. Since $3.7 = \frac{37}{10}$ and $.003 = 3/1,000$, $3.7 \times .003 = \frac{37}{10} \times 3/1,000 = 111/10,000$ or 111 ten-thousandths. Write 111 ten-thousandths as a decimal numeral.

.0111

38. From frame 37 we see that $3.7 \times .003 = .0111$. Notice that $37 \times 3 = 111$, so it is obvious from where the 111 comes. The only problem left is to determine how to know where to place the decimal point. In frame 37, the denominator of 10,000 came from multiplying the denominator of 10 (from $\frac{37}{10}$) by the denominator of 1,000 (from $3/1,000$). Thus, the placement of the decimal point is determined by the place value of the last digit in each decimal numeral. Since the 7 was in the tenths place and the 3 in the thousandths, the last digit of the answer will be in the $\frac{1}{10} \times 1/1,000 = 1/10,000$ or ten-thousandths place. In what place would the last digit go in $.13 \times .9$?

$$\frac{1}{100} \times \frac{1}{10} = \frac{1}{1,000}$$
or thousandths

39. A faster way of determining where the decimal point should be placed uses the fact that moving the decimal point one place to the left is the same as dividing by 10. For example, if the decimal point in 137. is moved one place, you obtain 13.7. $13.7 = 13\frac{7}{10} = \frac{137}{10} = 137 \div 10$. In the same way if you move the decimal point two places, it is the same as dividing by 100; three places is the same as dividing by 1,000; etc. Thus, moving the decimal point four places to the left is the same as dividing by _____.

10,000

40. Thus, in multiplying $3.7 \times .003$ we would move the decimal point of 111. four places to the left to divide by 10,000, getting .0111. In the numerals 3.7 and .003, how many digits are after the decimal point? Also, how many places was the decimal point moved to find the correct answer?

The answer is 4 to both questions.

41.
> When multiplying decimal numerals, multiply the decimals as though they were whole numbers; then move the decimal point to the left the same number of places as there are digits after the decimal point in the decimals being multiplied.

Example: $.041 \times 3.5$

$$
\begin{array}{r}
.041 \\
\times \quad 3.5 \\
\hline
205 \\
123 \quad \\
\hline
.1435 \\
\end{array}
$$

Find 31.2 × .04.

$$
\begin{array}{r}
1.248 \\[4pt]
31.2 \\
\times\ .04 \\
\hline
1.248
\end{array}
$$

42. Find 300 × 1.04.

$$
\begin{array}{r}
312 \\[4pt]
300. \\
\times\ 1.04 \\
\hline
1200 \\
000\ \ \\
300\ \ \ \ \\
\hline
312.00
\end{array}
$$

43. Find .17 × .48.

.0816

44. Find 6.875 × 3.1416.

21.5985

45. Consider the problem .1 × .049 = .0049. If both decimal numerals were rounded off or measured values, it is possible that .1 was actually .14 while .049 might have been .0494. In this case the product is .006916, not .0049. Since the first significant digits are different, the answer should be rounded off to one significant digit in this case. If you try other problems in this manner, you will find the following general rule:

> When multiplying rounded off decimal numerals the answer should be rounded off to the same number of significant digits as the decimal numeral with the fewest significant digits.

Example: Since in 1.2 × 30.3 1.2 has two significant digits and 30.3 has three significant digits, the answer should be rounded off to two significant digits.
Find 1.2 × 30.3 and round off the answer to two significant digits.

3.6
(3.636 is 3.6 rounded off to two significant digits.)

46. Multiply and round off to the correct number of significant digits:

40.3 × .017

.69

47. Multiply and round off to the correct number of significant digits:

3,200 × .241

770

48. Multiply and round off to the correct number of significant digits:

.061 × .134

.0082

Exercise Set 2-4.   Work all the problems before checking your answers.

1. Multiply (do not round off answers)
    (a) 400 × .003
    (b) .317 × .037
    (c) 4.81 × 41.6

    _____
    _____
    _____

2. Multiply and round off to the correct number of significant digits
    (a) 5100 × .123
    (b) .017 × .003
    (c) 4.125 × 3.14

    _____
    _____
    _____

2-5 DIVISION OF DECIMALS

If 9.25 feet of round stock weighs 22.2 pounds, how much does 1 foot weigh? To solve this problem you must divide 22.2 by 9.25. In this section you will learn a technique for dividing decimals.

49. $2.3 \div .07 = \dfrac{2.3}{.07}$

If .07 is multiplied by 100, then 2.3 must be multiplied by 100 also. We then find

$$\frac{2.3 \times 100}{.07 \times 100} = \frac{230}{7}$$

(which we know how to solve) since both decimal points are moved two places to the right if multiplied by 100. We can thus divide decimals simply by first moving the decimal points the required number of places to make the divisor a whole number. How far would the decimal points have to be moved in 4.3 ÷ 2.165?

Three places, since the decimal point in the divisor, 2.165, must be moved three places in order to make it a whole number.

50.

> To divide decimals, set up the problem as a long division problem and move both decimal points to the right the necessary number of places to make the divisor a whole number. Place the decimal point in the quotient at the spot it has been moved to and then divide as with whole numbers.

Example: .32 ÷ 1.6

(a) $1.6\overline{).32}$

Move the decimal to the right the necessary number of spaces.

(b)
```
      .2
16 ) 3.2
     32
```

Put decimal point in quotient and divide.

Divide 6.40 ÷ .16   40 (answer)
```
      40.
.16 ) 6.40
```

51. When you move the decimal point, if there are not enough digits, zeros must be added.

Example: 2.4 ÷ .173
```
.173 ) 2.400
```
(Add two zeros.)

Divide 7 ÷ .1.
(*Note*: The decimal point in 7 is after the 7 since it is a whole number.)

70 (answer)
```
      70.
.1 ) 7.0
```

52. **Example:**   .21 ÷ 7
Since 7 is already a whole number, the decimal point does not have to be moved. Thus, the decimal point moves straight up.

$$7 \overline{).21}$$

After the decimal point each place must have a digit in it; that is, there must be a digit above the 2 and the 1.

How often does 7 divide evenly into 2?   0 times

53. Since 7 divides evenly into 2 zero times, a 0 is placed above the 2. Finish the problem below.
```
      .0
7 ) .21
    0
    21
```

```
      .03
7 ) .21
    0
    21
    21
```

54. Divide (enter zeros as needed) .1353 ÷ 4.1.

.033 (answer)

$$
\begin{array}{r}
.033 \\
4.1\overline{)\,.1353} \\
\underline{123} \\
123 \\
\underline{123} \\
\end{array}
$$

55. Divide (enter zeros as needed) 7.5 ÷ 1.25.

6 (answer)

$$
\begin{array}{r}
6. \\
1.25\overline{)\,7.50} \\
\underline{750} \\
\end{array}
$$

56. Quite often the division of decimals will result in a division that will never have a remainder of 0. In such a case you have two options—you may stop the division at a certain point and write the remainder over the divisor and reduce to lowest terms or you may round off the decimal at a certain point. The former option will be considered first.

Example: Divide to the hundredths place and then write the remainder as a fraction:

$$1.1 \div .3$$

$$
\begin{array}{r}
3.66\tfrac{2}{3} \\
.3\overline{)\,1.100} \\
\underline{9} \\
20 \\
\underline{18} \\
20 \\
\underline{18} \\
2 \\
\end{array}
$$

(0's are added in order to continue the division)
Write remainder (2) over divisor (3).

Thus, $1.1 \div .3 = 3.66\tfrac{2}{3}$.

Divide .65 ÷ .48 to the hundredths and write the remainder as a fraction.

$1.35\tfrac{5}{12}$ (answer)

$$
\begin{array}{r}
1.35\tfrac{20}{48} \\
.48\overline{)\,.6500} \\
\underline{48} \\
170 \\
\underline{144} \\
260 \\
\underline{240} \\
20 \\
\end{array}
$$

$\left(\tfrac{20}{48} = \tfrac{5}{12}\right)$

57. Divide 4.1 ÷ 21.5 to the thousandths and write the remainder as a fraction.

$$
\begin{array}{r}
.190\frac{30}{43} \\
21.5\overline{)4.1000} \\
215 \\
\overline{1950} \\
1935 \\
\overline{150}
\end{array}
$$

58. **Example:** Divide 4.2 ÷ 3.1 and round off to the nearest tenth. In order to round off we must know what the digit after the digit in the tenths place is. For this reason you must divide to the hundredths and then use this number to round off the answer to the nearest tenth.

$$
\begin{array}{r}
1.35 \\
3.1\overline{)4.200} \\
31 \\
\overline{110} \\
93 \\
\overline{170} \\
155
\end{array}
$$

Round off 1.35 to the nearest tenth.

1.4 (Thus, 4.2 ÷ 3.1 = 1.4 to the nearest tenth.)

59. Divide 4.8 ÷ .415 and round off to two significant digits.

12 (answer)

$$
\begin{array}{r}
11.5 \\
.415\overline{)4.800} \\
415 \\
\overline{650} \\
415 \\
\overline{2350} \\
2075
\end{array}
$$

60.
> The rule for rounding off the answer to a division problem in which the decimal numerals are rounded off or measured values is the same as for multiplication—round off the answer to the same number of significant digits as are in the numeral with the fewest significant digits.

Divide .437 ÷ 6.1 and round off to the correct number of significant digits.

.072 (answer)

$$
\begin{array}{r}
.0716 \\
6.1\overline{)43700} \\
427 \\
\overline{100} \\
61 \\
\overline{390} \\
366
\end{array}
$$

.0716 rounded off to two significant digits equals .072.

61. Divide 4.1 ÷ .003 and round off to the correct number of significant digits.

1,000 (answer)

$$
\begin{array}{r}
1366. \\
.003\,\overline{)4.100} \\
3 \\
\hline
11 \\
9 \\
\hline
20 \\
18 \\
\hline
20 \\
18 \\
\hline
\end{array}
$$

Exercise Set 2-5.   Work all the problems before checking your answers.

1. Divide and round off to the hundredths place.
   (a) 4 ÷ .031
   (b) .41 ÷ 3
   (c) 2.7 ÷ 3.1
   (d) .0541 ÷ .27

2. Divide and round off to the correct number of significant digits.
   (a) 3.1 ÷ .04
   (b) .073 ÷ 34
   (c) 21 ÷ 7.1
   (d) .3041 ÷ 2.041

2-6   CHANGING FRACTIONS TO DECIMALS

Suppose the blueprint you are using indicates that a certain measurement should be $\frac{1}{8}''$. If you use an outside micrometer to check the piece, the reading will be in decimal notation. What reading would you like to find? In order to know you must change $\frac{1}{8}''$ to a decimal. This ability will also be needed when working with percents. For example, if $\frac{1}{12}$ of the parts were defective in the first week and 7% of the parts were defective in the second week, has the company reduced the percent of parts that are defective? To compare $\frac{1}{12}$ and 7%, it would be easiest to change $\frac{1}{12}$ to a percent, which requires that you first change $\frac{1}{12}$ to a decimal.

---

62. The fraction $\frac{1}{3}$ is equivalent to 1 ÷ 3 because the fraction bar is equivalent to division. Write $\frac{7}{9}$ as a division problem.

7 ÷ 9

63. Since $\frac{4}{5} = 4 \div 5$, you can change $\frac{4}{5}$ to a decimal numeral by dividing 4 by 5. Change $\frac{4}{5}$ to a decimal numeral.

.8 (answer)

$$
\begin{array}{r}
.8 \\
5\,\overline{)4.0} \\
40 \\
\hline
0
\end{array}
$$

64. **Example:** Change $\frac{2}{3}$ to a decimal numeral and round off the answer to the nearest tenth.

$$\begin{array}{r} .66 \\ 3\overline{)2.0} \\ \underline{18} \\ 20 \\ \underline{18} \end{array}$$ answer: .7
(rounded off to the nearest tenth)

65. Change $\frac{15}{13}$ to a decimal numeral rounded off to the nearest hundredth.   1.15

66. Change $\frac{17}{11}$ to a decimal numeral rounded off to the nearest thousandth.   1.545

Exercise Set 2-6.   Work all the problems before checking your answers.

1. Change to a decimal numeral.
   (a) $\frac{9}{8}$ _____
   (b) $\frac{3}{16}$ _____

2. Change to a decimal numeral rounded off to the nearest thousandth.
   (a) $\frac{16}{13}$ _____
   (b) $\frac{5}{9}$ _____
   (c) $\frac{1}{7}$ _____

SELF-TEST

1. What is the place value of the indicated digit in 8,617.04253?
   (a) 7   _____
   (b) 6   _____
   (c) 5   _____
   (d) 0   _____
   (e) 4   _____

2. Add
   (a) $\begin{array}{r} .7 \\ .3 \\ + .2 \\ \hline \end{array}$

   (b) $7 + .08 + 1.345$ _____

3. Subtract
   (a) $7 - 4.36$ _____
   (b) $\begin{array}{r} 6.08 \\ - 1.99 \\ \hline \end{array}$

   (c) $3.1 - .078$ _____

4. Round off as indicated:
   (a) 6849 to the nearest hundred.
   (b) 71.09 to the nearest one.
   (c) .3497 to the nearest thousandth.
   (d) 40.0746 to three significant digits.

5. Assuming that each numeral is a measured value, round off the following problems to the correct number of significant digits.
   (a) .049 ÷ 3.27
   (b) 5400 × .003
   (c) 6 + 2.087
   (d) 6.00 − 2.344

6. Multiply (do not round off the answer)
   (a) .312 × 41.7
   (b) 30 × .0031

7. Divide (do not round off the answer)
   (a) 7 ÷ .05
   (b) .0065 ÷ 130

8. Write $\frac{11}{9}$ as a decimal rounded off to the nearest hundredth.

APPLIED PROBLEMS

A. If the bore diameter is 3.8765 inches, name the digit with the following place value.

	Place Value	Digit
1.	tenth	
2.	ones	
3.	ten-thousandths	
4.	hundredths	
5.	thousandths	

B. Write the following micrometer readings in words.

	Reading	Stated in Words
1.	1.03	
2.	3.445	
3.	2.045	
4.	.031	
5.	1.034	

C. Round off the following main journal diameter measurements to three significant digits.

	Measured Value	Rounded off Value
1.	2.2983″	
2.	2.44095″	
3.	2.74810″	
4.	2.74499″	
5.	2.2615″	

D. Round off the following horsepower ratings to the nearest one.

	Ratings	Rounded off Ratings
1.	192.47	
2.	78.063	
3.	208.9	
4.	20.407	
5.	99.86	

E. The total current, W, in a certain circuit is found by adding the individual currents X, Y, and Z. Find the missing values in the table below.

	W	X	Y	Z
1.	8.612 amps		4.07 amps	2.384 amps
2.		.824 amp	1.625 amps	2.07 amps
3.	7.92 amps	3.8 amps		2.107 amps
4.	8 amps	2.003 amps	1.84 amps	
5.		3.1 amps	2.07 amps	1.85 amps

Example: 1. Since $8.612 = X + 4.07 + 2.384$, $8.612 = X + 6.454$ (adding $4.07 + 2.384$). Since subtraction is the opposite of addition, $X = 8.612 - 6.454 = 2.158$ amps.

F. If the cylinder bore of X inches is rebored Y inches, the new bore is Z inches. Find the missing values in the table below.

	X	Y	Z
1.	3.375		3.400
2.	3.625	.020	
3.	3.970		4.000
4.		.025	3.75
5.	3.609		3.625

G. If a collar has an outside diameter of X inches and an inside diameter of Y inches, it is Z inches thick. Find the missing values in the table below.

	X	Y	Z
1.	1.500	1.228	
2.	2.025		.14
3.	1.875	1.75	
4.		1.75	.382
5.	1.903		.275

H. If your salesman drove W miles last week in a car which gets X miles per gallon of gasoline which costs Y cents per gallon, he spent Z dollars for gasoline. Find the missing values in the table below.

	W	X	Y	Z
1.	924	19.2	62.9	
2.		18.7	59.9	25.45
3.	820	21.7		16.80
4.	645		60.9	19.64
5.	845	17.8	61.9	
6.		18.4	58.9	20.41

Example: 1. Since the car gets 19.2 miles per gallon, the number of gallons used is

$$\frac{924 \text{ miles}}{19.2 \text{ miles/gallon}} = 48.125 \text{ gallons}$$

Also, 62.9 cents per gallon = .629 dollars per gallon, so (48.125 gallons)(.629 dollars/gallon) = $30.30 (rounded off to the correct number of significant digits).

I. The total gear reduction, $X:1$, of an automobile is found by multiplying the transmission ratio, $Y:1$, by the rear axle ratio, $Z:1$.

	X	Y	Z
1.	6.3	2.1	
2.	6.0		2.6
3.	6.4	2.2	
4.	5.6		3.2
5.		1.8	2.9

Example: 1. Since $6.3 = 2.1 \times Z$, $Z = 6.3/2.1 = 3$.

J. The surface area of a piston is found by multiplying the bore times itself times .785. Find the area for each bore given below.

	Bore	*Area*
1.	3.875″	
2.	9.8 cm	
3.	3.75″	
4.	8.4 cm	
5.	4.01″	

K. After a voltage drop of X volts, the initial voltage of Y volts is reduced to Z volts.

	X	Y	Z
1.	2.8		112.2
2.		221.4	216.3
3.	3.17		108.2
4.		115.3	111.73
5.	4.32	114.0	

L. The current of X amps splits at a certain point into Y amps and Z amps.

	X	Y	Z
1.	2.0	.81	
2.	1.68		1.09
3.		3.1	.6
4.	3.05	2.1	
5.	1.49		.6
6.		1.03	.4

Example: 1. The sum of Y and Z must be X, so
$Z = X - Y = 2 - .81 = 1.19$ amps.

M. Find the difference between the specifications for the part and the micrometer reading in each case.

	Specification	Micrometer Reading	Difference
1.	3.75″	3.8″	
2.	.211″	.199″	
3.	2.603″	2.58″	
4.	3.000″	2.876″	
5.	6.18″	6.312″	

N. If you have ever pushed a car, you probably noticed that it took much more effort to get the car moving than to keep it moving once it is moving. This property is described by saying that the *coefficient of friction* is higher when the car is stopped than it is while it is in motion. The coefficient of friction, c.f., is a fraction that represents the ratio of the force, f, required to start an object moving or keep it moving to the weight, w, of the object, so c.f. $= f/w$ or c.f. $\times w = f$. Find the missing dimension in the table. Round answers that are not exact to the nearest hundredth.

	c.f.	f	w
1.	.1		2,500 lb
2.	.07		3,200 lb
3.		275	3,000 lb
4.		300	2,750 lb
5.	.06		3,125 lb
6.		226	2,225 lb

Example: 1. $(.1) \times 2,500 = f$, so $250 = f$.

O. The *feed* on a lathe is the amount a cut advances along the piece on each revolution. Thus, if a piece is moved forward .01″ with each revolution, the feed is .01 inch per revolution or just .01″. If the feed is .01″, a cut $10(.01″) = .10″$ long will be made in 10 revolutions. How long a cut will be made in X revolutions with a feed of $Y″$?

	X	Y	Cut
1.	35	.007	
2.	78	.011	
3.	47	.009	
4.	60	.015	
5.	43	.008	

3

measurement

OBJECTIVES

After completing this chapter the student should be able to

1. Define the terms *degrees*, *minutes*, and *seconds*.
2. Find the sum of angular measurements.
3. Find the difference of two angular measurements.
4. Read a standard micrometer.
5. Read a vernier caliper.
6. Read a bevel protractor.

PRETEST 3   (1–53)

(1–22) **1.** (a) $85' = $ _____ degrees _____ minutes

  _____ ; _____

(b) Find $75°2'' - 41°45'10''$.

_____

(23–33) **2.** What is the measurement indicated on the portion of a micrometer shown in Fig. 3-1?

_____

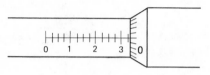

FIGURE 3-1

(34–53) **3.** What is the reading on a vernier if the 0 on the vernier proper lies between 1.275″ and 1.300″ and the vernier proper has 25 units in .6″ while the main scale has 24 units in .6″ assuming that the eleventh mark on the vernier proper lines up with a mark on the main scale?

_____

3-1   ANGULAR MEASUREMENT

Thousands of years ago by studying the location of formations of stars man knew that the position of stars in the sky changed from day to day and that eventually they returned to their original positions. Since they did not have accurate means of measuring the positions, they could only approximate the time lapse until a star returned to its original position. It was commonly accepted that the time lapse was 360 days for many years, although we now know that it is a little over 365 days, the length of a year. Since primitive man assumed that the earth was the center of the universe and everything revolved around it, it was natural for them to break up the revolution about a point into 360 parts—each part corresponding to one day's change. This led to the concept of a degree in measuring angles.

---

1. In essence, primitive man divided the circle into 360 equal parts. We shall do the same in defining a measure for angles called a *degree*.
In Fig. 3-2 a line has been drawn from the center of the circle to the

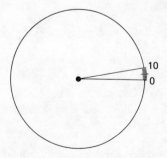

FIGURE 3-2

70

starting point, 0. A second line from the center terminates at a point that is 10 equal units, called *degrees*, away from the original point, 0. These two lines form an angle. The measure of an angle, how large it is, is defined as the number of equal units or degrees on the circle between the first and second line. Thus, since there are 10 equal units between the two lines, the measure of the angle is _____ degrees.

10

2. Note that any circle with center at the point where the two lines that form the angle meet can be used in determining the measure of the angle without changing the number of units encompassed by the angle—only the size of the equal units will change in conjunction with the change in the size of the circle. Thus, a degree is $\frac{1}{360}$ of the angle (at the center of any circle) traversed in making one revolution around the circle. The symbol ° is used as an abbreviation for degrees. Thus, 3 degrees is written 3°. Write 42 degrees using the symbol °.

42°

3. As technology advanced and it became necessary to have measurements more accurate than to the nearest degree, each degree was divided into 60 equal parts. Just as in telling time, each part is called a *minute*. Thus, it takes $60 \times 360 =$ _____ minutes to make one revolution.

21,600

4. The symbol for a minute is ′, so 10 minutes is written 10′. How would you write 7 minutes?

7′

5. Since $1' = \frac{1}{60}°$,
$70' = \frac{70}{60}° = 1\frac{10}{60}° = 1°10'$
$75' = 1$ degree _____ minutes

15

6. $115' =$ _____ degree _____ minutes

1; 55

7. $1° = 60'$ so
$2° = 2(60') = 120'$ and
$3° =$ _____ minutes.

180

8. $\frac{3}{4}° =$ _____ minutes

$\frac{3}{4}(60) = 45$

9. $1\frac{1}{2}° =$ _____ minutes

90

10. In the same way, each minute may be divided into 60 equal parts. Each of these parts is called a *second*. There are
$(60)(21,600) =$ _____ seconds in one revolution.

1,296,000

11. The symbol used to abbreviate second is ″. Thus, 7 seconds is written 7″. How would you write 2 minutes 13 seconds?

2′13″

12. How would you write 5 degrees 6 seconds?   5°6″

13. Since $1'' = \frac{1}{60}'$,
    $73'' = \frac{73}{60}' = 1'13''$.
    $66'' = $ _____ minute _____ seconds   1;   6

14. $130'' = $ _____ minutes _____ seconds   2;   10

15. $150'' = $ _____ minutes _____ seconds   2;   30

16. Since $1' = 60''$,
    $3' = $ _____ seconds. $3(60) = 180$

17. $\frac{1'}{2} = $ _____ seconds   30

18. It will be necessary for you to be able to add and subtract angular measurements later in this text. As with decimals you can only add and subtract like units, so some borrowing may be necessary and you will have to line up seconds with seconds, etc.

    Example: Find $3°14'' + 13°4'53''$.

    $$\begin{array}{r} 3°0'14'' \\ + \; 13°4'53'' \\ \hline 16°4'67'' \end{array}$$

    A zero may be added for any missing units. Then add like units.

    Since $67''$ is larger than one minute, it should be changed to minutes and seconds.
    $67'' = 1'7''$
    Thus, $16°4'67'' = 16°4' + 1'7'' = 16°5'7''$.
    Find $4°43' + 27'6''$.

         $$\begin{array}{r} 4°43'0'' \\ + \; 0°27'6'' \\ \hline 4°70'6'' \end{array}$$
    $70' = 1°10'$, so
    $4°70'6'' = 5°10'6''$.

19. Find $37°40'12'' + 49°10'45''$. $86°50'57''$

20. **Example:** Find $180° - 79°4''$.

    (a) $$\begin{array}{r} 180°0'0'' \\ - \;\; 79°0'4'' \\ \hline \end{array}$$

    Set up problem, adding zeros as needed.

(b)   179°60′0″
   −   79° 0′4″

Borrow 1° from 180° since you cannot subtract 4″ from 0″.
1° = 60′

(c)   179°59′60″
   −   79° 0′ 4″
      100°59′56″

Borrow 1′ from 60′.
1′ = 60″.   Then subtract.

Find 75°4″ − 39′12″.

$$\begin{array}{r} 59' \\ 74°\cancel{60'}64'' \\ \cancel{75°}\ \cancel{0'}\ \cancel{4''} \\ -\ \ 0°39'12'' \\ \hline 74°20'52'' \end{array}$$

21.   Find 4°3′13″ − 2°17′5″.

$$\begin{array}{r} 3°63' \\ \cancel{4°}\ \cancel{3'}13'' \\ -\ 2°17'\ 5'' \\ \hline 1°46'\ 8'' \end{array}$$

22.   Find 90° − 25°34″.

$$\begin{array}{r} 90°\ 0'\ 0'' \\ -\ 25°\ 0'34'' \\ \hline 64°59'26'' \end{array}$$

Exercise Set 3-1.   Work all the problems before checking your answers.

1.   Define the term *degree*. _____
   _____

2.   75′ = _____ degrees _____ minutes   _____ ; _____

3.   1° = _____ minutes   _____

4.   2′ = _____ seconds   _____

5.   245″ = _____ minutes _____ seconds   _____ ; _____

6.   Add
   (a) 45°25′ + 35°35″   _____
   (b) 16′45″ + 53′49″   _____

7.   Subtract
   (a) 39°45′40″ − 21°39′21″   _____
   (b) 100° − 49°41′52″   _____
   (c) 45°5″ − 21°5′45″   _____

3-2 THE MICROMETER

The greatest accuracy possible using a steel rule is $\frac{1}{64}$ inch. Quite often, as in bore measurements, however, we want accuracy to several thousandths of an inch. In order to obtain measurements with that kind of accuracy, you must use a micrometer. Figure 3-3 shows an outside micrometer used for measuring outer dimensions such as the diameter of a fitting. A micrometer is used when the desired measurement should be accurate to the thousandths of an inch. Automotive technicians use an inside micrometer to measure such things as the bore of a cylinder, while the machinist often checks the accuracy of his work using a micrometer.

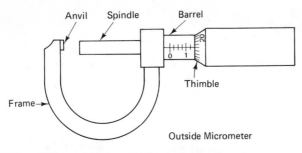

FIGURE 3-3

23. Refer to Fig. 3-3. In using the outside micrometer we measure the distance between the anvil and the spindle. To measure the length of the object in Fig. 3-4 the anvil would be placed on one side of the object and the spindle would be extended until it meets the opposite side of the object. This is done by turning the thimble, which turns the screw threads inside the micrometer, thus lengthening the exposed portion of the spindle. To understand how this is done, consider some screw threads. As the screw threads are turned, a small distance is moved by the spindle, which is attached to the screw threads. This is because with each revolution the screw moves the distance between two adjacent threads. This distance is called the *pitch*. Since the pitch is usually very small, a micrometer can be very accurate. For example, if the pitch is .025″, what distance will be moved in $\frac{1}{5}$ of a revolution?

.005″
$\frac{1}{5} \times .025″ = .005″$

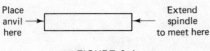

FIGURE 3-4

24. If the pitch is .025″, how far will the spindle move in $\frac{1}{25}$ of a revolution?

.001″
$\frac{1}{25} \times .025″ = .001″$

25. Since if we make $\frac{1}{25}$ of a revolution on a screw with a pitch of .025″, the spindle moves .001″, the micrometer will give us measurements to

the thousandths of an inch, if we can indicate each $\frac{1}{25}$ of a revolution. Refer to Fig. 3-3. When the thimble makes one complete revolution, the screw has made one complete revolution also. Thus, if we can mark each $\frac{1}{25}$ of a revolution on the thimble, we shall have the desired accuracy. How many equally spaced marks would have to be placed on the thimble to indicate each $\frac{1}{25}$ revolution?

25, since then each unit distance will be 1 part out of the 25 that make up one complete revolution.

26. Refer to Fig. 3-3. The 25 marks on the thimble are made adjacent to the barrel of the micrometer. In addition, on the barrel are markings to indicate each revolution of the screw threads. Thus each marking on the barrel indicates _____ inch.

.025, since the pitch is .025″.

27. Every fourth mark on the barrel is given a number starting with 1. Four revolutions is equal to 4 × .025″ = _____ inch.

.100

28. Refer to Fig. 3-5. From frame 26 it is obvious why every fourth mark on the barrel is marked since they indicate .100″, .200″, etc. On the barrel the 3 indicates _____.

.300″

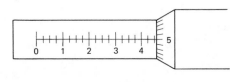

FIGURE 3-5

29. In Fig. 3-5 the 4 indicates .400″, but notice that 2 more revolutions were made beyond the 4. These 2 revolutions add on an additional _____ inch. (Recall that 1 revolution = .025″.)

.050″
2 × .025″ = .050″

30. Refer to Fig. 3-5. Notice on the thimble that the screw has also been turned $\frac{5}{25}$ of a revolution. Since $\frac{1}{25}$ revolution equals .001″, this indicates an additional _____ inch in the measurement.

.005

31. The total measurement in Fig. 3-5 is .400″ + .050″ + .005″ = .455″. What is the measurement indicated in Fig. 3-6?

.370″
.300″ + .050″ + .20″ = .370″

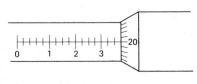

FIGURE 3-6

32. When finding an inside measurement such as bore, an inside micrometer is used. The size of the extension rod varies depending on the size to be measured. Measurements between 2″ and 3″ use a 2–3 inch extension rod, those between 3″ and 4″ use a 3–4 inch extension rod, etc. Thus, for a measurement of 6.851 inches you would use the 6–7 inch rod. What size rod would you use for a measurement of 5.698″?

5–6 inch rod

33. The inside micrometer is read in exactly the same way as the outside micrometer, except that the starting point of the outside micrometer is 0, while the starting measure point of the inside micrometer is 2″ or 3″ or 4″, etc., depending on the size extension rod used. What is the reading in Fig. 3-7 if the rod size is 3–4 inches?

3.146″
3.000″ + .100″ + .025″ + .021″
                              = 3.146″
or,
 rod + barrel + thimble
                  = total

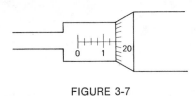

FIGURE 3-7

Exercise Set 3-2.   Work all the problems before checking your answers.

1. Each mark on the barrel of a micrometer represents _____ inch.

2. What is the measurement indicated on the micrometer in Fig. 3-8?

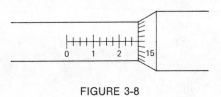

FIGURE 3-8

3-3 THE VERNIER

A vernier scale is often used to allow for a more precise reading. If a vernier scale is used with a micrometer, measurements accurate to ten-thousandths of an inch are possible. In this section you will learn the concept behind this scale, which was invented by Pierre Vernier in 1637.

34. Refer to Fig. 3-9. In scale A the inch has been divided into four equal units, so each unit is $\frac{1}{4}$ inch or .25″. In scale B the inch has been divided into five equal units, so each unit is $\frac{1}{5}$ inch or .2″. What is the difference between the $\frac{1}{5}$″ marking and the $\frac{1}{4}$″ marking?

$\frac{1}{4}$″ − $\frac{1}{5}$″ = $\frac{5}{20}$″ − $\frac{4}{20}$″
= $\frac{1}{20}$″ or .05″

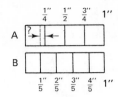

FIGURE 3-9

35. Refer to Fig. 3-9. If scale B is moved until the $\frac{1}{5}''$ mark lines up with the $\frac{1}{4}''$ line, how far has scale B been moved?

Since the distance between $\frac{1}{4}''$ and $\frac{1}{5}''$ is $\frac{1}{20}''$, scale B has been moved $\frac{1}{20}''$ or .05".

36. Refer to Fig. 3-9. What is the distance between $\frac{1}{2}''$ and $\frac{2}{5}''$?

$\frac{1}{2}'' - \frac{2}{5}'' = \frac{5}{10}'' - \frac{4}{10}''$
$= \frac{1}{10}''$ or .10"

37. If scale B was moved until the $\frac{3}{5}''$ mark lined up with the $\frac{3}{4}''$ mark, how far would scale B have to be moved?

$\frac{3}{20}''$ or .15", the distance between $\frac{3}{4}''$ and $\frac{3}{5}''$.

38. How far would scale B have to be moved so that the $\frac{4}{5}''$ mark lined up with the 1" mark on scale A?

$1'' - \frac{4}{5}'' = \frac{1}{5}'' (= \frac{4}{20}''$ or .20")

39. Refer to frames 33–38. Notice that scale B must be moved $\frac{1}{20}''$ or .05" for the first mark on scale B to line up with the first mark on scale A. It must be moved $\frac{2}{20}''$ or .10" for the second mark to line up, $\frac{3}{20}''$ or .15" for the third mark to line up, and $\frac{4}{20}''$ or .20" for the fourth mark to line up, where $\frac{1}{20}''$ or .05" is the difference between the smallest unit on scale A and the smallest unit on scale B. *In general, the use of a second scale with one more or one less smaller equal units in the same length as the first scale will allow measurements accurate to the difference between the lengths of the smaller units on the two scales. This is the principle behind the vernier.* If there are four equal units in .1 inch on the *main scale* and five equal units in .1 inch on the *vernier proper*, what is the smallest measurement possible?

Since there are four units in .1", each unit on the main scale is .1" ÷ 4 = .025", while each unit on the vernier proper is .1" ÷ 5 = .020". The smallest measurement possible is .025" − .020" = .005".

40. What is the smallest possible measurement if there are 24 units in .6" on the main scale and 25 units in .6" on the vernier proper?

.6" ÷ 24 = .025"
.6" ÷ 25 = .024"
.025" − .024" = <u>.001"</u>

41. The reading on the main scale of the vernier caliper in Fig. 3-10 is between 1.25″ and 1.50″ since the 0 on the vernier proper is between 1.25″ and 1.50″. Since the second mark on the vernier proper lines up with a mark on the main scale, the caliper reading is 2(.05″) or .10″ beyond the 1.25″ reading since the smallest measurement possible in this case is .05″. Thus, the reading is 1.25″ + .10″ = _____.

1.35″

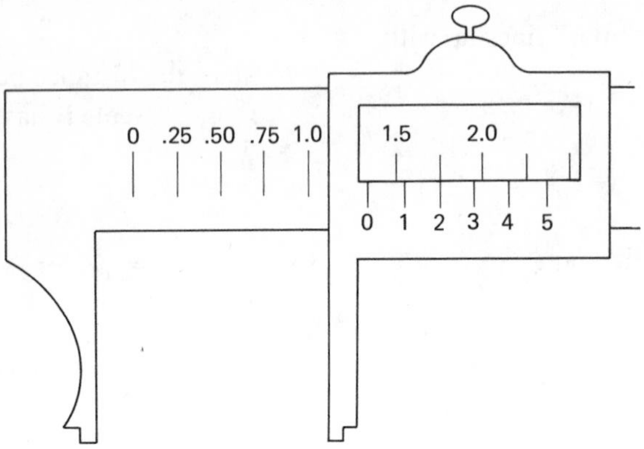

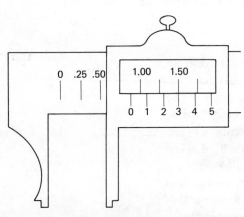

FIGURE 3-10

42. What is the reading on the vernier caliper in Fig. 3-11?

.75″ + .15″ = .90″
[The 0 is between .75″ and 1.00″; the third mark lines up so you must add 3(.05″) to .75″.]

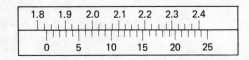

FIGURE 3-11

43. On the portion of a vernier caliper in Fig. 3-12, the smallest possible measurement is .001″. What is the measurement indicated on the vernier caliper?

1.825″ + .006″ = 1.831″

FIGURE 3-12

44. The next few frames will deal with the *bevel protractor*, a tool for measuring angles. The basic bevel protractor will measure angles to the nearest degree. By using the concept of the vernier, however, smaller measurements can be made. Recall that if each degree is divided into 60 equal units, this smaller unit is called a *minute*.

Thus,

1 degree = 60 minutes

2 degrees = 2°(60) = 120 minutes

23 degrees = _____ minutes

$23°(60) = 1.380$

45. The smallest unit on the main scale of a bevel protractor is 1 degree. On the vernier scale there are 12 equal units in 23 degrees or 1,380 minutes. What is the value of each of these units on the vernier scale?

$\dfrac{1,380}{12} = 115$ minutes

46. Refer to Fig. 3-13. Is the first mark on the vernier scale closer to one degree or two degrees?

$2°$

FIGURE 3-13

47. Thus, to determine the smallest possible measurement with the vernier-type bevel protractor, we should subtract the value of each unit on the vernier scale from 2°. What is the smallest measurement possible with this bevel protractor?

5 minutes

2° = 120 minutes

120 minutes − 115 minutes = 5 minutes

48. Thus, if the third mark on the vernier lines up with a mark on the main scale, you must add 3(5 minutes) = 15 minutes (written 15′). Refer to Fig. 3-14. The main scale is composed of four portions, which go from 0° to 90° (90° to 0°), while the vernier has two parts, which go from 0′ to 60′ in steps of 5 minutes. There are two places on the main scale where the reading is 0°. These points can be used as a reference when using the vernier. Notice that the reading at the 0 on the vernier in Fig. 3-14 is between 86° and 87° to the right of the 0° mark. Also one of the marks on each part of the vernier scale lines up. Since the eleventh mark on the left lines up, this reading is 11(5′) = _____.

55′

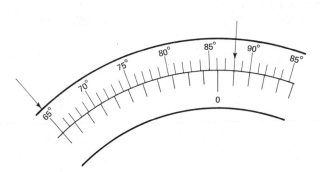

FIGURE 3-14

49. Refer to Fig. 3-14. The first mark on the right half of the vernier scale lines up so this reading is _____.

5′

50. Refer to Fig. 3-14. If we add 55′ to 86°, we find 86° 55′ or almost 87°; while if we add 5′ to 86°, we find 86° 5′, which is very close to 86°. From the figure, which reading do you think is correct?

86° 5′, since the reading on the main scale is just over 86°.

51. As you saw in frame 49, the correct reading is obtained by adding the vernier reading on the right half of the scale to the lower reading on the main scale.

In general,
  (a) If the reading on the main scale is to the left of the 0° mark, add the reading on the left half of the vernier scale to the smaller reading on the main scale.
  (b) If the reading on the main scale is to the right of the 0° mark, add the reading on the right half of the vernier scale to the smaller reading on the main scale.

Which half of the vernier scale would you use for the reading in Fig. 3-15?

The left half, since the reading is to the left of the 0° mark.

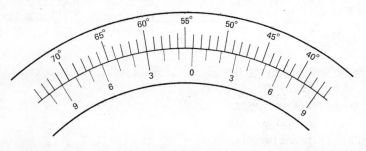

FIGURE 3-15

52. Refer to Fig. 3-15. The reading on the main scale at the 0 point of the vernier scale is between _____ and _____.

54°; 55°

53. Refer to Fig. 3-15. What is the reading on the bevel protractor since the third mark on the left half of the vernier lines up?

54° + 3(5′) = 54°15′

Exercise Set 3-3.   Work all the problems before checking your answers.

1. What is the smallest possible measurement if there are five units in .12″ on the main scale and six units in .12″ on the vernier proper?

2. What is the reading on a vernier, if
   (a) The 0 lies between 3.125″ and 3.150″.

(b) On the main scale there are 24 equal units in .6″, while on the vernier proper there are 25 units in .6″, and

(c) The fourth mark on the vernier proper lines up. _____

3. What is the reading on a vernier bevel protractor if the 0 is between 42° and 43° to the right of 0° and the seventh mark on the left-hand side and fifth mark on the right-hand side line up? _____

SELF-TEST

1. Define the terms *degree*, *minute*, and *second*. _____ twerl five _____

_____

_____

_____   _____ ; _____

2. 138′ = __2__ degrees __18__ minutes

3. 2° = __120__ minutes

4. 10′ = __600__ seconds

5. 10°13′5″ + 7°49′7″ = $8°2′12″$

6. 98°59′17″ + 11°48″ = $99°0′5″$

7. 65°49′ − 46°35′ = $19°14′$

8. 102° − 75°41″ = $26°59′19″$

9. Each mark on the barrel of a micrometer represents __.025__ inch.

10. What is the measurement indicated on the micrometer in Fig. 3-16? __.271__

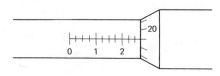

FIGURE 3-16

11. What is the measurement in Fig. 3-16 if it is an inside micrometer with a 3–4 inch extension rod?   3.271

12. What is the smallest possible measurement using a vernier, if there are 10 units in .11 inch on the main scale and 11 units in .11 inch on the vernier proper? _____

13. Using the vernier described in Problem 12, what is the reading if the 0 on the vernier proper lies between .022″ and .033″ and the seventh mark on the vernier proper lines up? _____

14. What is the reading on a vernier bevel protractor if the 0 is between 31° and 32° to the left of the 0° mark and the eighth mark on the left-hand side and the fourth mark on the right-hand side line up? _____

APPLIED PROBLEMS

A. If the measure of one angle is X degrees and the measure of another is
    Y degrees, with $X + Y = 180°$, the two angles are called *supplementary*
    angles. Find the missing values in the table below.

	X	Y
1.	15°	
2.		28°30′
3.	100°39″	
4.		49°50′35″
5.	156°59′7″	

B. Since the sum of the angles of a triangle is 180°, $X + Y + Z = 180°$
    where X, Y, and Z are the measures of the three angles of a triangle.
    Find the missing values in the table below.

	X	Y	Z
1.	21°	151°	
2.		27°7′	39°55′
3.	49°7″		68°4′
4.	61°21′23″	102°13′49″	
5.		49°49′27″	1°10′40″

C. In turning a piece of work on a lathe the machinist must set the compound
    rest at an angle, X, to the center line of the work that is equal to one-half

	X	Y
1.	30°	
2.		30°
3.	22°30′	
4.		90°
5.		55°

the included angle, *Y*, of the desired piece. Find the missing values in the table below.

D. The measure of each interior (inside) angle of a regular polygon (one in which all sides are of the same length) with *n* sides can be found from the following: $180° - (360° \div n)$. In the table below find the measure of each interior angle, *A*, of a regular polygon of *n* sides to the nearest minute.

	n	*A*
1.	3	
2.	5	
3.	10	
4.	11	
5.	13	

E. Using the fact that the sum of the interior angles of a triangle is 180° and Fig. 3-17 as a guide, derive the formula $(n - 2) \times 180°$ for the sum of the interior angles of a polygon of *n* sides. [*Hints*: (a) How many triangles can be formed by connecting vertices of an *n*-sided polygon in such a way that no angle of one triangle is included in an angle of another triangle? (b) What is the sum of the angles of this many triangles? (c) What is the relationship between this sum and the sum of the interior angles?]

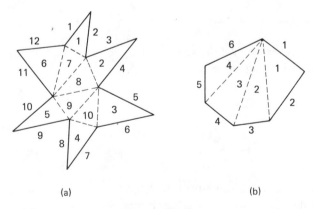

(a)   (b)

FIGURE 3-17

F. Use a micrometer to measure the thickness of a page of the text, a pencil, and a wire. Have your instructor check the measurements.

G. Use a bevel protractor to measure each of the interior angles in Fig. 3-17(b).

H. Using the formula derived in Problem E, find the sum of the interior angles of each *n*-sided polygon in the table below.

	n	*Sum*
1.	5	
2.	4	
3.	7	
4.	12	
5.	10	

I. The sum of the interior angles formula can be used to find an interior angle in a polygon if all the other angles are known. Find the missing angle below for an *n*-sided polygon with the angles given.

	n	*Given Angles*	*Missing Angle*
1.	4	40°, 50°, 120°	
2.	5	60°, 45°, 55°, 50°	
3.	6	30°, 35°, 70°, 50°, 20°15′	
4.	7	30°, 80°, 70°, 15°, 25°, 52°7″	
5.	8	100°, 21°10′, 39°10′10″, 30°10′5″, 40°, 50°, 2°13″	

J. Two angles *A* and *B* are said to be *complementary* if the sum of their measures is 90°. Find the complement of *A* in each case below.
1. 16°
2. 15°5′
3. 85°41″
4. 35°4′3″
5. 47′23″

K. In order to determine the area of a tract of land, you must first determine the interior angles. The surveyor will usually indicate bearings but not specify the angles involved. Thus, in order to determine the angles you must be able to comprehend bearing readings. In general, a bearing tells you how many degrees east or west of the north-south line a certain boundary line runs. For example, a reading of N 30° E means that when facing north the line makes an angle of 30° to the east of the north-south line, while a reading of S 40° W means that when facing south the line makes an angle of 40° to the west of the north-south line. In Fig. 3-18, find all the interior angles. The first one has been done for you. It is suggested that you set up the N-S, E-W system as shown to assist you and extend lines as needed. Note that the N-S and E-W lines form an angle of 90° so complementary angles are often utilized.

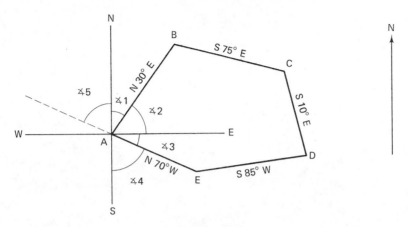

FIGURE 3-18

Note first that ∡1 is 30° since line *AB* is 30° to the east of the N-S line. Thus, since ∡2 is its complement, ∡2 = 60°. ∡3 is the complement of ∡4 so if we can determine ∡4, we can find ∡3. Note that ∡4 and ∡5 have the same measure. Since the line is 70° to the west of the N-S line, ∡5 is 70°. Thus, ∡4 is 70° so ∡3 is 20°, the complement of ∡4. Thus, the interior angle, which is ∡2 and ∡3, measures 60° + 20° = 80°.

4

signed numbers

OBJECTIVES

After completing this chapter the student should be able to
1. Find the opposite of any signed number.
2. Find the absolute value of any signed number.
3. Locate any signed number on the number line and give the signed number represented by a given point on the number line.
4. Find the sum of two signed numbers.
5. Find the difference of two signed numbers.
6. Find the product of two signed numbers.
7. Find the quotient of two signed numbers.

(1–18) **1.** (a) What is the opposite of ⁻7?

_____

(b) What is the absolute value of ⁺6?

_____

(19–23) **2.** Label ⁻8 on the number line in Fig. 4-1.

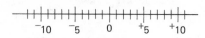

   ⁻10 ⁻5 0 ⁺5 ⁺10

FIGURE 4-1

(24–43) **3.** Find

(a) ⁻7 + ⁻5

_____

(b) 6 + ⁻13

_____

(44–52) **4.** Find

(a) ⁻6 − 4

_____

(b) 7 − ⁻4

_____

(c) ⁻10 − ⁻3

_____

(d) 6 − 15

_____

(53–77) **5.** Find

(a) ⁻19 · 4

_____

(b) $\dfrac{^-81}{^-3}$

_____

(c) 48 ÷ ⁻16

_____

4-1 DEFINITIONS

In the real world you must often deal with opposites: up–down, above–below, left–right, and more–less. Since mathematics is used to interpret the world, there must be opposites in mathematics also. For example, if you represent the temperature of seventeen degrees above zero by 17°, how can you represent seventeen degrees below zero. If you represent a profit of one thousand five hundred dollars by $1500, how do you represent a loss of seven hundred fifty dollars? One possible answer to these problems is to use signed numbers, or integers as they are sometimes called.

---

1. Seven degrees below zero is the opposite of seven degrees above zero. Since they do not have the same value, you cannot use the same value, you cannot use the same symbol for the two readings. A method is needed to distinguish between a temperature above zero and one below zero. One way of doing this is to assign one sign to those numbers above zero and another sign to those numbers below zero. The usual notation is to call readings above zero _positive_ and assign a "⁺" sign to them and call readings below zero _negative_ and

assign them a "⁻" sign. Thus, 12° above zero is written ⁺12°, while
12° below zero is written _____.

⁻12°

2. The signed number "⁻7" is read "negative seven," while "⁺6" is read
"positive six." How would you read "⁻13"?

negative thirteen

3. Using signed numbers, a thermometer might look like the one in
Fig. 4-2. The reading at the point A is ⁻3°. What is the reading at
the point B?

⁻9°

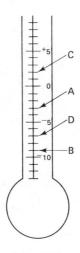

FIGURE 4-2

4. What are the readings at points C and D in Fig. 4-2?

C: ⁺2°
D: ⁻7°

5. The signed number ⁻6 is the opposite of ⁺6. What is the opposite of
⁺13?

⁻13

6. What is the opposite of ⁻12?

⁺12

7. What is the opposite of ⁻2?

⁺2

8. If a number does not have a sign, such as 6, it is assumed to be
positive. Thus, 6 = ⁺6. What is the opposite of 6?

⁻6

9. Is 9 positive or negative?

positive

10. What is the opposite of ⁻3?

⁺3 (or 3)

11. How many spaces is ⁺7° from 0 on the thermometer in Fig. 4-2?

seven spaces

12. How many spaces is ⁻7° from 0 on the thermometer in Fig. 4-2?   seven spaces

13. The 7 in ⁺7° and ⁻7° refers to the number of spaces the number is
     from 0°, while the sign tells whether it is above or below 0°. This
     idea of the distance from zero is important enough to be given a
     name; it is called the *absolute value* of the number. Thus, the
     absolute value of ⁺7 is 7 and the absolute value of ⁻7 is also 7.
     What is the absolute value of ⁺3?   3

14. What is the absolute value of ⁻6?   6

15. What is the absolute value of 4 (= ⁺4)?   4

16. What is the absolute value of 0?   0

17. What is the absolute value of 13?   13

18. What is the absolute value of ⁻14?   14

Exercise Set 4-1.   Work all the problems before checking your answers.

1. How would you read the following numerals?
    (a) ⁺6   _____
    (b) ⁻4   _____

2. What is the opposite of each numeral?
    (a) ⁺6   _____
    (b) ⁻5   _____
    (c) 16   _____

3. What is the absolute value of each numeral?
    (a) ⁺17   _____
    (b) 0   _____
    (c) ⁻41   _____

4-2 THE NUMBER LINE

If the thermometer in Fig. 4-2 is turned on its side, we see what is commonly
called a *number line*. A number line is useful in understanding the addition,
subtraction, multiplication, and division of signed numbers.

---

19. A number line is a straight line with marks equally spaced on the
     line. Figure 4-3 is a number line. Starting by labeling one mark 0,

the marks to the right of the 0 point are labeled successively $^+1$, $^+2$, $^+3$, $^+4$, etc. The marks to the left of 0 are labeled successively $^-1$, $^-2$, $^-3$, $^-4$, etc. Each point represents a signed number. By going far enough to the right or left of 0 we can find a point to represent any signed number we desire. Mark the point that represents the signed number $^-4$ on the number line in Fig. 4-3.

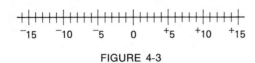

FIGURE 4-3

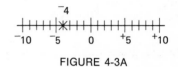

FIGURE 4-3A

20. Mark $^-7$ on the number line in Fig. 4-3.

FIGURE 4-4A

21. Mark $^+3$ and $^-13$ on the number line in Fig. 4-3.

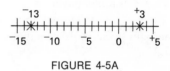

FIGURE 4-5A

22. What number is represented by the X on the number line in Fig. 4-6.

$^-11$

FIGURE 4-6

23. What number is represented by the points A and B in Fig. 4-6?

A: $^-12$
B: $^+6$ (or 6)

Exercise Set 4-2. Work all the problems before checking your answers.

1. Label the following points on the number line in Fig. 4-7.
   (a) $^-6$
   (b) 0
   (c) $^+14$
   (d) $^-4$

2. What signed number is represented by each of the points A, B, and C in Fig. 4-7?

A _____
B _____
C _____

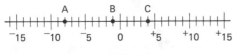

FIGURE 4-7

4-3 ADDITION OF SIGNED NUMBERS

"John's business made a profit of $1,700 during the first quarter but had a loss of $2,100 during the second quarter. What is his profit or loss for the first half of the year?" This problem and other similar problems can be solved easily by knowing how to add signed numbers. If we use "$^+$" for a profit and "$^-$" for a loss, this problem can be stated as $^+1,700 + {}^-2,100$. This section deals with solving addition problems of this type.

---

24. Recall that on the number line $^+$ means to *go to the right*," while $^-$ means *go to the left*," since the positive numbers lie to the right of 0 and the negative numbers lie to the left of 0. We can use the number line to demonstrate the addition of signed numbers. For example, the simple problem $^+2 + {}^+3$ is shown on the number line in Fig. 4-8 as a movement of two spaces to the right from 0 (since $+$ in $^+2$ means *go to the right*) followed by a movement of three spaces to the right (for $^+3$). Notice that you end on the point $^+5$, which is the answer.
    Indicate the addition problem $^+3 + {}^+4$ on the number line in Fig. 4-8.

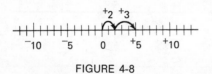

FIGURE 4-8

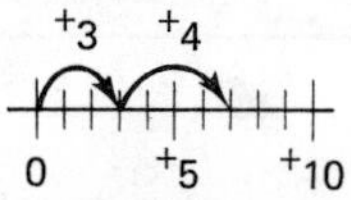

FIGURE 4-8A

$+7$ (answer)

25. Since "$^-$" means "go to the left," $^-3 + {}^-1$ can be indicated by a movement of three spaces to the left followed by a movement of one space to the left.
    Thus, $^-3 + {}^-1 =$ _____.

$^-4$

26. Indicate $^-4 + {}^-2$ on a number line.

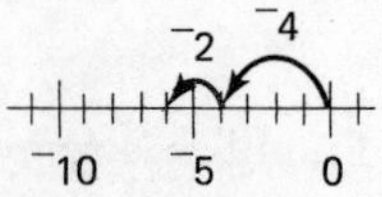

FIGURE 4-9A

27. From frame 26, $^-4 + {}^-2 =$ _____.

$^-6$

28. Indicate $^-7 + {}^-3$ on a number line and find the answer to the problem.

FIGURE 4-10A

$^-7 + {}^-3 = {}^-10$

29. Refer to frames 24–28. Notice that when both signs are the same in an addition problem, the movements on the number line are in the same direction. Thus, the answer is farther from 0 than either of the numbers being added. In fact, the absolute value or distance from 0 of the answer is equal to the sum of the absolute values of the numbers being added. For example, you saw that $^-7 + {}^-3 = {}^-10$. The absolute value of $^-7$ is 7, the absolute value of $^-3$ is 3, while the absolute value of $^-10$ is _____, the sum of 7 and 3.

10

30. The findings of frame 29 result in the following rule:

When adding signed numbers with the same sign, add the absolute values of the numbers being added and use the same sign to find the answer.

Thus, in adding $^-14 + {}^-13$, add $14 + 13$ to obtain 27 and then use the negative sign. Therefore, $^-14 + {}^-13 = {}^-27$. Find $^-21 + {}^-16$.

$^-21 + {}^-16 = {}^-37$

31. Find $^-7 + {}^-36$.

$^-43$

32. Consider the problem $^-4 + {}^+5$. Since $^-$ means go to the left, while $^+$ means go to the right, the addition problem can be indicated as shown in Fig. 4-11.
Thus $^-4 + {}^+5 = $ _____.

$^+1$

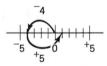

FIGURE 4-11

33. Indicate $^+7 + {}^-10$ on a number line.

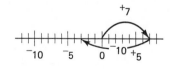

FIGURE 4-12A

34. From Fig. 4-12A in frame 33 we see $^+7 + {}^-10 = $ _____.

$^-3$

35. Notice on the number line in Fig. 4-12A that since the signs of the numbers are opposite, when indicated on the number line they go in opposite or different directions. What is the difference between the absolute values of the numbers being added in $^+7 + {}^-10$?

absolute values: 7, 10
difference: $10 - 7 = 3$

36. You saw that $^+7 + {}^-10 = {}^-3$. You also saw that the difference in the absolute values of $^+7$ and $^-10$ is 3. Which has the largest absolute value, $^+7$ or $^-10$?

$^-10$

37. Refer to frames 35 and 36. $^-10$ has the largest absolute value and the sign of the answer is the same as the sign of this number. If you consider the number line in Fig. 4-12A, you will see why this is true. Since you had to go to the left farther than you had to go to the right, the answer will be on the left or negative side. This leads to the following rule:

> When adding signed numbers with different signs, find the difference of the absolute values of the numbers being added and use the sign of the number with the largest absolute value.

Find $^-6 + 5$. (*Note*: $5 = {}^+5$)

absolute values: 6, 5
difference: $6 - 5 = 1$
Since $^-6$ is larger in absolute
value, the sign is negative. Thus,
$^-6 + 5 = {}^-1$

38. Find $^-6 + 9$.

3

39. Find $^-6 + {}^-2$.

$^-8$ (Note that the signs are the same.)

40. Find $^-9 + 17$.

8

41. Find $^-6 + 3$.

$^-3$

42. Find $^-13 + {}^-42$.

$^-55$

43. Find $^-9 + 10$.

1

Exercise Set 4-3.   Work all the problems before checking your answers.
Find the following sums.

1.   ⁻6 + ⁻3   _____

2.   6 + ⁻7   _____

3.   ⁻9 + 13   _____

4.   ⁻11 + ⁻36   _____

5.   ⁻15 + ⁻15   _____

6.   ⁻19 + 24   _____

4-4   SUBTRACTION OF SIGNED NUMBERS

"The temperature at 6:00 a.m. was 6°. During the day the temperature dropped 9°.
What is the final temperature?" This problem can be represented by the
subtraction problem 6 − 9. This section treats the solution of such problems.

44.   Subtraction is the opposite operation of addition. The problem
      ⁺5 + ⁺2 is represented by a movement to the right of five spaces
      followed by a movement to the right of two spaces to find the
      answer ⁺7. The answer to ⁺5 − ⁺2 is ⁺3, however, not ⁺7. Since
      subtraction is the opposite of addition, after a movement of five
      spaces to the right, instead of going two spaces more to the right for
      ⁺2, you must do the opposite and go two spaces to the left. The
      diagram is shown in Fig. 4-13.
      Indicate 6 − 4 on the number line in Fig. 4-13.

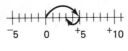

FIGURE 4-13

FIGURE 4-13A

45.   Refer to the number lines in Figs. 4-13 & 4-13A. Compare the
      number line for ⁺5 − ⁺2 with the one in Fig. 4-14A for ⁺5 + ⁻2.
      They are exactly the same. Indicate the addition problem 6 + ⁻4 on
      the number line in Fig. 4-14 and compare it with the one for 6 − 4.

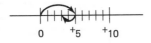

FIGURE 4-14

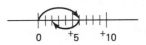

FIGURE 4-14A

(They are exactly the same.)

46. Refer to frames 44 and 45. Since the number line representations are the same for the two problems, the answers must be the same. Thus, $^+5 - {}^+2 = {}^+5 + {}^-2$ and $6 - 4 = 6 + {}^-4$. Notice that $^-2$ is the opposite of $^+2$, just as $^-4$ is the opposite of 4, while addition $(+)$ is the opposite of subtraction $(-)$. This leads to the following rule for subtracting signed numbers:

When subtracting signed numbers
(a) Change the number being subtracted to its opposite and the minus sign $(-)$ to a plus sign $(+)$.
(b) Then solve the resulting addition problem to find the answer.

Example: Find $3 - 5$.
   (a) Change $3 - 5$ to an addition problem:

$$3 - 5 = 3 + {}^-5$$

   (b) Solve the resulting addition problem.

$$3 + {}^-5 = {}^-2$$

Find $7 - 19$.

(a) $7 - 19 = 7 + {}^-19$
(b) $7 + {}^-19 = {}^-12$

47. The opposite of $^-4$ is $^+4$. Thus, $6 - {}^-4 = 6 + {}^+4 = 10$. Find $7 - {}^-2$.

$7 - {}^-2 = 7 + {}^+2 = 9$

48. Find $^-4 - {}^-3$.

$^-4 - {}^-3 = {}^-4 + {}^+3 = {}^-1$

49. Find $^-7 - 6$.

$^-7 - 6 = {}^-7 + {}^-6 = {}^-13$

50. Find $7 - 11$.

$7 - 11 = 7 + {}^-11 = {}^-4$

51. Find $^-3 - {}^-17$.

$^-3 - {}^-17 = {}^-3 + {}^+17 = 14$

52. Find $^-11 - 6$.

$^-11 - 6 = {}^-11 + {}^-6 = {}^-17$

Exercise Set 4-4. Work all the problems before checking your answers.

1. Find $6 - 14$.

2. Find $^-4 - {}^-7$.

3. Find $6 - {}^-4$.

4. Find $^-5 - 13$.

5. Find $^-9 - {}^-2$.

4-5 MULTIPLICATION AND DIVISION OF SIGNED NUMBERS

As you have seen, signed numbers can be used to solve various problems. Their greatest value to you, however, will come in solving equations in Chapter 5. This section deals with multiplication and division of signed numbers. They are treated together because, as you saw with fractions, every division problem can be changed to a multiplication problem. Thus, the rules for division will be the same as the rules for multiplication.

---

53. Multiplication is a simplified form of repeated addition. For example, 3×7 is an abbreviation for $7 + 7 + 7$, adding 7 three times. Write 4×9 as an addition problem.

$4 \times 9 = 9 + 9 + 9 + 9$

54. Write $3 \times {}^-6$ as an addition problem.

$3 \times {}^-6 = {}^-6 + {}^-6 + {}^-6$

55. Since $3 \times {}^-6 = {}^-6 + {}^-6 + {}^-6$, $3 \times {}^-6 =$ _____.

${}^-18$, since ${}^-6 + {}^-6 + {}^-6 = {}^-18$.

56. Write $4 \times {}^-8$ as an addition problem and find its value.

$4 \times {}^-8 = {}^-8 + {}^-8 + {}^-8 + {}^-8$
$= {}^-32$

57. $7 \times 3 = 3 \times 7$ by the commutative law for multiplication (see Chapter 0). In the same way ${}^-7 \times 3 = 3 \times {}^-7$. Thus ${}^-7 \times 3 = 3 \times {}^-7 = {}^-7 + {}^-7 + {}^-7 =$ _____.

${}^-21$

58. Notice that the problems in frames 55–57 all involved one positive and one negative number. Also notice that the answer was always negative. Since this will always be true, we have the following rule:

> When multiplying signed numbers with different signs, the answer is the negative of the product of the absolute values of the signed numbers.

Thus, ${}^-7 \times 9 = {}^-63$ since $7 \times 9 = 63$ and we know the answer is negative.
Find $6 \times {}^-11$.

${}^-66$ since $6 \times 11 = 66$ and the signs are opposite.

59. $6 \times {}^-5$ can also be written as $6 \cdot {}^-5$ or $(6)({}^-5)$. Both the dot, $\cdot$, and the use of parentheses without a sign between the parentheses indicate multiplication. Write ${}^-7 \times 5$ in two other ways.

${}^-7 \cdot 5$
$({}^-7)(5)$

60. Find ${}^-7 \cdot 5$.

${}^-35$

61. Find $6 \cdot {}^-5$.

${}^-30$

62. Find $(^-13)(3)$. $^-39$

63. Find $(14)(^-7)$. $^-98$

64. Since $^+3 \times {}^-7$ means *add $^-7$ three times*, $^-3 \times {}^-7$ cannot mean the
     same thing. Because the 3 is negative rather than positive, $^-3 \times {}^-7$
     can be interpreted as *subtract $^-7$ three times*. Using this definition
     $^-3 \times {}^-7 = -{}^-7 - {}^-7 - {}^-7$. Whenever you subtract signed
     numbers, however, you change the number being subtracted to its
     opposite and add. Thus,

$$^-3 \times {}^-7 = -{}^-7 - {}^-7 - {}^-7$$
$$= +{}^+7 + {}^+7 + {}^+7$$
$$= \underline{\hspace{1.5cm}} \qquad\qquad {}^+21$$

65. As indicated in frame 64, the product of two negative signed numbers
     is a positive number. Since the product of two positive numbers is
     also positive, we know that the product of two signed numbers with
     the same sign is positive. Combining this with what we learned about
     the product of signed numbers with unlike signs, we find

     ┌───┐
     │ To multiply two signed numbers,   │
     │ (a)   Find the product of the absolute values of the numbers.   │
     │ (b)   If the signs of the numbers are *different*, the answer is   │
     │   *negative*; while if the signs are the *same*, the answer is *positive*. │
     └───┘

     Thus, $^-5 \cdot {}^-4 = {}^+20$ because the signs are the same.
     Find $(^-9)(^-7)$. $^+63$ or 63

66. Find $^-7 \cdot 4$. $^-28$

67. Find $^-17 \cdot {}^-5$. $^+85$ or 85

68. Find $^-6 \cdot {}^-13$. 78

69. Find $^-7(7)$. $^-49$

70. Find $19(6)$. 114

71. Find $^-8 \times {}^-7$. 56

72. Since $12 = \frac{12}{1}$ and $6 = \frac{6}{1}$, $12 \div 6$ can be written as $\frac{12}{1} \div \frac{6}{1}$. From
     Chapter 1 we know that this can be rewritten as $\frac{12}{1} \times \frac{1}{6}$. Thus, the
     division problem $12 \div 6 = \frac{12}{1} \times \frac{1}{6}$. In the same way,
     $^-14 \div 7 = {}^-\frac{14}{1} \times \frac{1}{7}$. Since this is the product of numbers with
     different signs, the answer is negative. Thus, $^-14 \div 7 = {}^-2$. Since
     every division problem can be changed to a multiplication problem,
     the rule for determining the sign must be the same as in

multiplication. Thus,

> To divide signed numbers,
> (a) Divide the absolute values.
> (b) If the signs are *different*, the answer is *negative*; while if the signs are the *same*, the answer is *positive*.

Find $^-28 \div {}^-7$. $^+4$ since the signs are the same.

73.   Find $^-24 \div 6$. $^-4$

74.   Find $84 \div {}^-7$. $^-12$

75.   Find $^-75 \div {}^-15$.   5

76. $\dfrac{65}{^-13} = 65 \div {}^-13$.

      Find $\dfrac{65}{^-13}$. $^-5$

77.   Find $\dfrac{^-98}{^-14}$. $^+7$

Exercise Set 4-5.   Work all the problems before checking your answers.

1.   Find $^-7 \cdot {}^-6$.   _____

2.   Find $^-9 \times 8$ _____

3.   Find $^-\frac{120}{8}$.   _____

4.   Find $^-91 \div {}^-7$ _____

5.   Find $(14)(^-6)$.   _____

6.   Find $\frac{49}{^-7}$.   _____

SELF-TEST

1. Find the opposite of each number below.
    (a) $^-6$ _____
    (b) 5   _____
    (c) $^+3$ _____

2. Find the absolute value of each number below.
    (a) $^+6$ _____
    (b) 0   _____
    (c) $^-7$ _____

3. How would you read $^-13$?   _____

4. Label ⁻7 and 8 on the number line in Fig. 4-15.

FIGURE 4-15

5. What signed number is represented by each of the points A and B in Fig. 4-15?

A _____

B _____

6. Find
   (a) ⁻7 + ⁻9
   (b) 13 + ⁻16
   (c) ⁻9 + 14

   _____

   _____

   _____

7. Find
   (a) ⁻7 − 4
   (b) ⁻6 − ⁻3
   (c) 7 − 13
   (d) 7 − ⁻5

   _____

   _____

   _____

   _____

8. Find
   (a) 7 · ⁻12
   (b) (⁻8)(⁻6)
   (c) ⁻3 × 8
   (d) $\dfrac{81}{⁻27}$
   (e) ⁻50 ÷ ⁻5
   (f) ⁻39 ÷ 3

   _____

   _____

   _____

   _____

   _____

   _____

APPLIED PROBLEMS

A. Find the following sums:

	Problem	Sum
1.	⁻3 + 2 + ⁻6	
2.	6 + ⁻3 + ⁻2	
3.	5 + 3 + ⁻7	
4.	⁻6 + ⁻4 + ⁻2	
5.	2 + ⁻5 + 3	
6.	⁻4 + 5 + ⁻2	

Example: 1. ⁻3 + 2 + ⁻6 = (⁻3 + 2) + ⁻6 = ⁻1 + ⁻6 = ⁻7

B. Find the following differences:

	Problem	Difference
1.	$^-3 - 2 - ^-4$	
2.	$6 - 3 - 5$	
3.	$^-5 - ^-3 - ^-4$	
4.	$7 - 6 - ^-2$	
5.	$^-7 - 3 - 6$	
6.	$^-2 - ^-4 - 6$	

Example: 1. $^-3 - 2 - ^-4 = ^-3 + ^-2 + 4 = (^-3 + ^-2) + 4 = ^-5 + 4 = ^-1$

C. Simplify the following expressions:

	Problem	Answer
1.	$6 - ^-3 + 4$	
2.	$^-5 + 6 - 8$	
3.	$^-2 - 5 + 3$	
4.	$6 - 9 + 2$	
5.	$8 + 1 - ^-4$	
6.	$^-6 - 3 + 2$	

Example: 1. $6 - ^-3 + 4 = 6 + 3 + 4 = (6 + 3) + 4 = 9 + 4 = 13$

D. Find the following products:

	Problem	Product
1.	$^-3 \cdot 4 \cdot ^-2$	
2.	$6 \times ^-3 \times 4$	
3.	$^-9 \cdot ^-2 \cdot ^-1$	
4.	$14 \times ^-5 \times ^-3$	
5.	$(^-6)(^-3)(^-7)$	
6.	$(2)(^-3)(13)$	

Example: 1. $^-3 \cdot 4 \cdot ^-2 = (^-3 \cdot 4) \cdot ^-2 = ^-12 \cdot ^-2 = 24$

E. Bob had receipts of X dollars and expenses of Y dollars for the day. What was his profit or loss, Z?

	X	Y	Z
1.	984.59	1,002.31	
2.	852.00	694.00	
3.	1,122.65	895.70	
4.	746.28	902.45	
5.	821.31	825.00	
6.	1,240.32	865.51	

Example: 1. Letting receipts be positive ($^+$) and expenses negative ($^-$), the sum, $984.59 + {}^-1,002.31$, is the profit or loss. Since the techniques for adding signed numbers apply to decimals, we take the difference of the absolute values $(1,002.31 - 984.59)$ and use the sign of the one with the largest absolute value.

$$984.59 + {}^-1,002.31 = {}^-17.72$$

Thus, there was a loss of $17.72.

F. Sam spent X dollars on new equipment that resulted in Y dollars in additional receipts. What was the additional (net) cost, Z, to him the first year for the new equipment?

	X	Y	Z
1.	1,000	650	
2.	1,200	450	
3.	950	750	
4.	2,500	2,100	
5.	3,000	1,225	
6.	1,700	1,530	

Example: 1. Let X be negative ($^-$) since it is an expense and Y be positive ($^+$) because it is money taken in. Then since $^-1,000 + 650 = {}^-350$, the additional cost is $350.

G. If Sam was allowed to subtract one-tenth (.1) the cost of the new equipment from his taxes, what was the net cost to him in Problem F?

Example: 1. In addition to adding X and Y we must now subtract .1 times X. Thus:

$$^-1,000 + 650 - (.1)(^-1,000) = -1,000 + 650 - \ ^-100$$
$$= (^-1,000 + 650) + 100$$
$$= -350 + 100 = \ ^-250$$

so the net cost was $250.

H. To change temperatures from the Fahrenheit scale to the Celsius scale, you must perform the following steps:
(a) Subtract 32 from the Fahrenheit temperature.
(b) Multiply this quantity by 5.
(c) Divide the product found in step (b) by 9.
Change the following Fahrenheit readings to the Celsius scale.

	Fahrenheit	*Celsius*
1.	14°	
2.	5°	
3.	−4°	
4.	212°	
5.	−22°	
6.	0°	

Example: 1. (a) $14 - 32 = 14 + \ ^-32 = \ ^-18$
  (b) $5 \times \ ^-18 = \ ^-90$
  (c) $^-90 \div 9 = \ ^-10$

Thus, the Celsius reading is $^-10°C$.

5

introduction to algebra

OBJECTIVES

After completing this chapter the student should be able to
1. Determine whether or not a specified letter is used as a variable.
2. Find the sum, difference, and product of monomials.
3. Change a complicated first-degree equation in one variable to an equivalent simpler equation.
4. Solve a first-degree equation in one variable by changing it to an equivalent equation that can be solved by inspection.
5. Solve first-degree equations involving fractions.
6. Solve a literal equation for a specified symbol.

(1–26) **1.** Simplify the following:

    (a) $8x - 9x$

    (b) $\dfrac{^-2}{3} \cdot \dfrac{6x}{7}$

_____

_____

(27–67) **2.** Solve for x;

$$4 - 2x + 3 = 4x$$

_____

(68–83) **3.** Solve for x:

$$\frac{5}{x} + \frac{2}{3} = \frac{2}{9}$$

_____

(84–90) **4.** Solve for x:

$$\frac{2}{x} + \frac{k}{x} = 7$$

_____

5-1 COMBINING POLYNOMIALS

What is a variable? In mathematics a **variable** is an object that can assume any of many possible values. In general, letters are used for variables. This concept is particularly helpful in solving many practical problems.

    In order to solve equations you will have to be able to work with terms involving variables. Such terms are called *monomials*. In this section we look at several operations on monomials.

---

1. For our purposes we shall define a variable as a letter which can be replaced by some number or value and still result in an expression which makes sense, without destroying the meaning of the original expression. For example, suppose the letter x refers to the *number* of people who attended the meeting. If we replace x by 5, we have the statement "5 people attended the meeting." This makes sense and we have not destroyed the original meaning (although the statement may be false). Thus x is a variable in this case. Is x a variable in "John won x letters in varsity sports."?

    Yes, since x can be replaced by 2 or 3, etc., and still make sense and not destroy the meaning.

2. Monomials follow the same general rules as numbers. Thus, since $5 + 5 + 5 = 3(5)$, $x + x + x = $ _____(x).

    3

3. Since $4(3) = 3 + 3 + 3 + 3$, $5x = $ _____ .

    $x + x + x + x + x$

4. Since $2x = x + x$, $2x + x = (x + x) + x = $ _____ $3x$

5. $3x + 2x = (x + x + x) + (x + x) = $ _____ $5x$

6. | To add (subtract) $6x$ and $7x$ we treat the x's like units and just add (subtract) the numbers—carrying along the x.

For example, 6 feet $+$ 7 feet $=$ 13 feet and $6x + 7x = $ _____. $13x$

7. $11x + 3x = $ $14x$

8. $5x + {}^{-}3x = $ $2x$

9. $9x + {}^{-}10x = $ ${}^{-}1x$

10. x is assumed to mean $1x$. Thus, $x + {}^{-}3x = 1x + {}^{-}3x = $ _____. ${}^{-}2x$

11. ${}^{-}5x + x = $ ${}^{-}4x$

12. $7x - 9x = (7 - 9)x = (7 + {}^{-}9)x = $ _____ $-2x$

13. $x - 10x = (1 - 10)x = (1 + {}^{-}10)x = $ _____ $-9x$

14. $12x - 13x = $ _____ ${}^{-}x$ (or ${}^{-}1x$)

15. $5x - 7x = 5x + {}^{-}7x = $ _____ ${}^{-}2x$

16. ${}^{-}6x - 6x = $ _____ ${}^{-}12x$

17. ${}^{-}5x - {}^{-}2x = ({}^{-}5 - {}^{-}2)x = ({}^{-}5 + 2)x = $ _____ ${}^{-}3x$

18. $6x - {}^{-}3x = $ _____ $9x$

19. 3 inches $+$ 3 inches $+$ 3 inches $+$ 3 inches $=$ 4(3 inches) $=$ 12 inches
    In the same way, $3x + 3x + 3x + 3x = 4(3x) = $ _____ $12x$

20. | To multiply a number times a monomial we simply multiply the number times the numerical portion of the monomial.

Thus, $5(2x) = (5 \cdot 2) \cdot x^* = 10x$. $4(6x) = $ _____ $(4 \cdot 6) \cdot x = 24x$

_____

*See the associative law in Introduction on p. 1.

21. $^-7(^-4b) =$ _____ $28b$

22. $\frac{1}{3}(3x) = (\frac{1}{3} \cdot \frac{3}{1})x = 1x$ or x

    $\frac{1}{4} \cdot (4x) =$ _____ $(\frac{1}{4} \cdot \frac{4}{1})x = 1x = x$

23. $\frac{1}{7}(7x) =$ _____ $(\frac{1}{7} \cdot \frac{7}{1})x = 1x = x$

24. $\frac{2}{3} \cdot (\frac{3}{2}x) = (\frac{2}{3} \cdot \frac{3}{2}) \cdot x = 1 \cdot x = x$

    $\frac{4}{5}(\frac{5}{4}x) = (\frac{4}{5} \cdot \frac{5}{4})x =$ _____ x

25. $\frac{3}{7}(\frac{7}{3}x) =$ _____ $(\frac{3}{7} \cdot \frac{7}{3})x = x$

26. $\dfrac{^-2}{3}\left(\dfrac{^-3}{2}x\right) =$ $\left(\dfrac{^-2}{3} \cdot \dfrac{^-3}{2}\right)x = x$

Exercise Set 5-1.   Work all the problems before checking your answers. Simplify each of the following:

1. $3x + 2x$ _____

2. $^-7x + ^-3x$ _____

3. $9x - 11x$ _____

4. $^-3x - ^-6x$ _____

5. $6x + ^-2x$ _____

6. $5(4x)$ _____

7. $^-3(2x)$ _____

8. $^-4(^-5x)$ _____

9. $\dfrac{^-1}{2}(^-2x)$ _____

10. $\dfrac{^-2}{3}\left(\dfrac{3}{5}x\right)$ _____

5-2 SOLVING FIRST-DEGREE EQUATIONS IN ONE VARIABLE

Consider Fig. 5-1. In Fig. 5-1(a) three x blocks plus two 1-pound blocks weigh the same as eleven 1-pound blocks. This can be stated by the equation $3x + 2 = 11$. If we take two 1-pound blocks from both sides of the scale, will the two sides still balance? Of course they will as long as we remove the same amount from both sides [see Fig. 5-1(b)]. In this case we find $3x = 9$ when we subtract 2 from both sides. This illustrates a basic law of equations—*the addition law for equations*—which states that the same quantity can be added (subtracted) from both sides of an equation without changing the solution. (Notice that if we remove different amounts from the two sides, the scale will not balance.) Now consider Fig. 5-1(c). One x block will weigh $\frac{1}{3}$ as much as three x blocks. Thus, one x block will weigh $\frac{1}{3}$ of 9 pounds or 3 pounds, which gives us the third diagram or $x = 3$. This illustrates the second basic law of equations, *the multiplication law for equations*, which states that both sides of an equation can be multiplied (divided) by the same number without changing the solution. These two laws will be used in this chapter to find simpler equivalent equations.

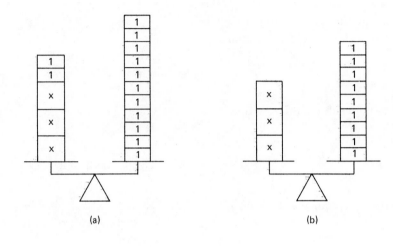

(a)   (b)

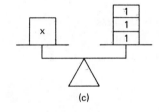

(c)

FIGURE 5-1

---

27. Consider the equations $3x + 2 = 11$ and $3x = 9$. Which equation could be solved more easily than the other?

$3x = 9$ because it is simpler.

28. The main difference between $3x + 2 = 11$ and $3x = 9$ is that the second equation has fewer terms since the 2 was eliminated from the left side. As noted above, the addition law for equations allows us to eliminate terms. In this case the 2 in $3x + 2 = 11$ can be eliminated by adding $^-2$ to both sides of the equation. What would you add to both sides of $2x + 5 = 9$ to eliminate the 5?

$^-5$

29. Notice that the 2 was eliminated by adding $^-2$ to both sides, while the 5 was eliminated by adding $^-5$. 5 and $^-5$ and 2 and $^-2$ are called _____ of each other.

opposites

30. From the frames above you can see that

> one way of simplifying an equation is to eliminate a term by adding its opposite to both sides of the equation to find a simpler equivalent equation.

In general, when solving equations you would like to have all the terms containing an x on one side and all the other terms on the opposite side. Thus, to simplify $2x = 4 + 3x$ you would try to eliminate the $3x$ on the right side of the equation in order to place all the x terms on the left side of the equation. How would you eliminate the $3x$?

Add ^-3x, its opposite, to both sides.

31. Simplify $2x = 4 + 3x$ by adding ^-3x to both sides of the equation.

$2x + (^-3x) = 4 + 3x + (^-3x)$
$-x = 4$

32. What is the solution to $-x = 4$?

$^-4$, since $- \; ^-4 = 4$.

33. Solve $5x - 3 = 7$ by first eliminating the $^-3$. (*Note*: $5x - 3 = 7$ is the same as $5x + \; ^-3 = 5$.)

$5x + \; ^-3 + (3) = 7 + (3)$
$5x = 10$
Solution: $x = 2$

34. The equation $5x = 3$ cannot be simplified by adding $^-5$ to both sides since $5x$ and $^-5$ are unlike terms and cannot be combined. To solve this equation you must use a different law, the multiplication law for equations, which states that if you multiply both sides of an equation by the same number, the resulting equation is equivalent to the original equation. Recall that to simplify $3x = 9$ we multiplied both sides of the equation by $\frac{1}{3}$. Note that 3 and $\frac{1}{3}$ are _____ of each other.

reciprocals

35. To solve the equation $5x = 3$, you would like to find an equivalent equation of the form x equals some number. To do this the 5 must be eliminated. Since $5x$ means "5 times x," the 5 can be eliminated by multiplying by its reciprocal as shown below:
$5x = 3$
$(\frac{1}{5})5x = (\frac{1}{5})3$
$x = $ _____

$(\frac{1}{5})(\frac{3}{1}) = \frac{3}{5}$

36. In frame 35 you saw that the solution to $5x = 3$ is $\frac{3}{5}$. What is the solution to $^-4x = 5$.
(*Hint*: Multiply by the reciprocal of $^-4$, $^-\frac{1}{4}$.)

$\left(\dfrac{^-1}{4}\right) \; ^-4x = \left(\dfrac{^-1}{4}\right)5$
$x = \dfrac{^-5}{4}$

37. Thus, to solve a first-degree equation the following steps are used:

> (a) Use the opposite to get all the numbers on one side of the equals sign and the x terms on the opposite side.
>
> (b) Multiply both sides by the reciprocal of the number in front of x.

Example: Solve $3x + 5 = 7$.
(a) $3x + 5 + (^-5) = 7 + (^-5)$
      $3x = 2$
(b) $(\frac{1}{3})3x = (\frac{1}{3})2$
      $x = \frac{2}{3}$

Solve $^-2x + 4 = 6$.

-1
(a) $^-2x + 4 + (^-4) = 6 + (^-4)$
      $^-2x = 2$
(b) $\left(\dfrac{^-1}{2}\right)^-2x = \left(\dfrac{^-1}{2}\right)2$
      $x = ^-1$

38. Since $3x - 7 = 3x + {}^-7$,
     $3x - 7 = 5$ is solved using the following steps.
(a) $3x - 7 = 5$
      $3x + {}^-7 + (7) = 5 + (7)$
      $3x = 12$
(b) $(\frac{1}{3})3x = (\frac{1}{3})12$
      $x = 4$

Solve $^-7x - 6 = {}^-13$.

(a) $^-7x + {}^-6 + (6) = {}^-13 + (6)$
      $^-7x = {}^-7$
(b) $\left(\dfrac{^-1}{7}\right)^-7x = \left(\dfrac{^-1}{7}\right)^-7$
      $x = 1$

39. To simplify $3x = 2x - 7$ we would like to have both x terms on the same side. Thus, if we eliminate the $2x$, we shall have the x terms all on the left side. The opposite of $2x$ is ^-2x. Using the addition law of equations we find
(a) $3x = 2x - 7$
    $3x + (^-2x) = 2x + (^-2x) - 7$
    $x = ^-7$

Solve $5x = 3x + 1$.

(a) $5x + (^-3x) = 3x + (^-3x) + 1$
      $2x = 1$
(b) $(\frac{1}{2})2x = (\frac{1}{2})1$
      $x = \frac{1}{2}$

40. Solve $8x = 3x - 2$.

(a) $8x + (^-3x) = 3x + (^-3x) - 2$
    $5x = ^-2$
(b) $(\frac{1}{5})5x = (\frac{1}{5})^-2$
    $x = \dfrac{^-2}{5}$

41. Solve $^-4x = 2x + 3$.

(a) $^-4x + (^-2x) = 2x + (^-2x)$
                          $+ 3$
    $^-6x = 3$
(b) $\left(\dfrac{^-1}{6}\right)^-6x = \left(\dfrac{^-1}{6}\right)3$
    $x = \dfrac{^-3}{6} = \dfrac{^-1}{2}$

42. To simplify $3 + 2x = 5x$, it would be simplest to eliminate the $2x$ since then the number 3 will be alone on one side and the x terms alone on the other side.
    (a) $3 + 2x + (^-2x) = 5x + (^-2x)$
        $3 = 3x$
    (b) $(\frac{1}{3})3 = (\frac{1}{3})3x$
        $1 = x$

    Solve $9x - 7 = 4x$.

(a) $9x + (^-9x) - 7 = 4x + (^-9x)$
    $^-7 = ^-5x$
(b) $\left(\dfrac{^-1}{5}\right)^-7 = \left(\dfrac{^-1}{5}\right)^-5x$
    $\frac{7}{5} = x$

43. $2 - 3x = 5x$ is the same as $2 + ^-3x = 5x$. What number would you have to add to both sides to eliminate the ^-3x?

$3x$

44. **Example:** Solve $2 - 3x = 5x$.
    (a) $2 + ^-3x + (3x) = 5x + (3x)$
        $2 = 8x$
    (b) $(\frac{1}{8})2 = (\frac{1}{8})8x$
        $\frac{1}{4} = x$

    Solve $7 - 5x = 2x$.

(a) $7 + ^-5x + (5x) = 2x + (5x)$
    $7 = 7x$
(b) $(\frac{1}{7})7 = (\frac{1}{7})7x$
    $1 = x$

45. To simplify $5 - 3x = 9$ you would like all the x terms on one side and all the numbers on the other side. We have the only x term on the left side but there are numbers on both sides. Since we want both

numbers on the side opposite the x term, which number should we eliminate?

5, since then the x term will be on the left and the numbers on the right side.

46. Solve $5 - 3x = 9$.

(a) $5 + (^-5) - 3x = 9 + (^-5)$
$^-3x = 4$

(b) $\left(\dfrac{^-1}{3}\right)^-3x = \left(\dfrac{^-1}{3}\right)4$

$x = \dfrac{^-4}{3}$

47. Solve $5x - 4 = 1$ using equivalent equations.

$5x + ^-4 + (4) = 1 + (4)$
$5x = 5$
$x = 1$

48. Solve $2x + 9 = 3$.

$2x + 9 + (^-9) = 3 + (^-9)$
$2x = ^-6$
$x = ^-3$

49. Solve $^-3x + 6 = ^-3$.

$^-3x + 6 + (^-6) = ^-3 + (^-6)$
$^-3x = ^-9$
$x = 3$

50. Solve $2x - 7 = 3$.

$2x - 7 + (7) = 3 + (7)$
$2x = 10$
$x = 5$

51. Solve $^-7x - 9 = 5$.

$^-7x - 9 + (9) = 5 + (9)$
$^-7x = 14$
$x = ^-2$

52. Solve $4x = 2x + 6$.

$4x + (^-2x) = 2x + (^-2x) + 6$
$2x = 6$
$x = 3$

53. Solve $^-6x = ^-3x + 9$.

$^-6x + (3x) = ^-3x + (3x) + 9$
$^-3x = 9$
$x = ^-3$

54. Solve $4x = ^-3x + 14$.

$4x + (3x) = ^-3x + (3x) + 14$
$7x = 14$
$x = 2$

55. Solve $2x + 7 = 3x - 1$.
    (*Hint*: Place all the x terms on one side and all numbers on the other side by using the addition law for equations.)

8 (answer)
Possible method:
$2x + (^-2x) + 7 = 3x + (^-2x) - 1$
$7 = x - 1$
$7 + (1) = x - 1 + (1)$
$8 = x$

56. Solve $^-3x - 7 = 2x + 3$.

$^-3x + (3x) - 7 = 2x + (3x) + 3$
$^-7 = 5x + 3$
$^-7 + (^-3) = 5x + 3 + (^-3)$
$^-10 = 5x$
$^-2 = x$

57. Solve $3 - 6x = 9$.

$3 + (^-3) - 6x = 9 + (^-3)$
$^-6x = 6$
$x = ^-1$

58. Solve $9 - 4x = ^-3$.

$9 + (^-9) - 4x = ^-3 + (^-9)$
$^-4x = ^-12$
$x = 3$

59. To solve $2x - 7x + 4 = ^-11$, note that $2x$ and ^-7x are like terms and thus can be combined.

    Solve $2x - 7x + 4 = ^-11$.

$^-5x + 4 = ^-11$
$^-5x + 4 + (^-4) = ^-11 + (^-4)$
$^-5x = ^-15$
$x = 3$

60. Solve $3 - 6x + 4 = 1$.

$7 - 6x = 1$
$7 + (^-7) - 6x = 1 + (^-7)$
$^-6x = ^-6$
$x = 1$

61. Solve $3x = ^-7$

$(\frac{1}{3})3x = (\frac{1}{3})^-7$
$x = \dfrac{^-7}{3}$

62. Solve $8x - 3 = ^-2$.

(a) $8x - 3 + (3) = ^-2 + (3)$
     $8x = 1$
(b) $(\frac{1}{8})8x = (\frac{1}{8})1$
     $x = \frac{1}{8}$

63. Solve $7 - 3x = 5$.

$7 + (^-7) - 3x = 5 + (^-7)$
$^-3x = ^-2$
$\left(\dfrac{^-1}{3}\right)^-3x = \left(\dfrac{^-1}{3}\right)^-2$
$x = \frac{2}{3}$

64. Solve $9 - 2x = 4x$.

$9 - 2x + (2x) = 4x + (2x)$
$9 = 6x$
$(\frac{1}{6})9 = (\frac{1}{6})6x$
$\frac{3}{2} = x$

65. Solve $^-3x + 2 = 5x$.

$^-3x + (3x) + 2 = 5x + (3x)$
$2 = 8x$
$(\frac{1}{8})2 = (\frac{1}{8})8x$
$\frac{1}{4} = x$

66. Solve $10x - 4 = 3x + 3$.

$10x + (^-3x) - 4 = 3x + (^-3x) + 3$
$7x - 4 = 3$
$7x - 4 + (4) = 3 + (4)$
$7x = 7$
$x = 1$

67. Solve $^-4x + 2 = 3x - 1$.

$^-4x + (^-3x) + 2 = 3x + (^-3x) - 1$
$^-7x + 2 = ^-1$
$^-7x + 2 + (^-2) = ^-1 + (^-2)$
$^-7x = ^-3$
$\left(\dfrac{^-1}{7}\right)^-7x = \left(\dfrac{^-1}{7}\right)^-3$
$x = \frac{3}{7}$

Exercise Set 5-2.   Work all the problems before checking your answers. Solve each of the following equations for x.

1. $2x - 5 = 7$

2. $3x + 4 = ^-2$

3. $^-6x + 1 = ^-5$

4. $^-x + 6 = ^-3$

5. $3x = 7$

6. $^-7x = ^-5$

7. $2 - 4x = 9 + x$

8. $7 + 2x = ^-3x - 1$

9. $^-7x + 3 - 4x = ^-8$

10. $2 - 6x + 3 = 2x$

5-3 SOLVING FRACTIONAL EQUATIONS IN ONE VARIABLE

Fractions always seem to cause problems when you have to work with them. In this section you will learn a method that eliminates the need to work with fractions in equations of the first degree.

---

68. We have already seen that if you multiply both sides of an equation by the same number, you get an equivalent equation (the multiplication law for equations). Consider $\frac{1}{3}x = 4$. The fraction $\frac{1}{3}$ sticks out in the equation. If we could eliminate the fraction, we could solve the equation as we have before. The way to do that is to multiply both sides of the equation by a number that will eliminate the denominator of the fraction. 3 works nicely because it is the reciprocal of $\frac{1}{3}$: $(3)\frac{1}{3}x = (3)4$ or $x = 12$. Eliminate the fraction in $\frac{1}{4}x = 7$.

$(4)\frac{1}{4}x = 7(4)$

$x = 28$

69. Eliminate the fraction in $\frac{1}{5}x = {}^-6$. (As a general rule a fraction within an equation can be eliminated by multiplying both sides of the equation by the denominator of the fraction.)

$(5)\frac{1}{5}x = (5)^-6$

$x = {}^-30$

70. When there is more than one term on a side, the distributive law* tells us we must multiply each term by the number by which we are multiplying both sides.

Example:

$\frac{1}{3}x + 6 = 9$

$3(\frac{1}{3}x + 6) = (3)9$

$(3)\frac{1}{3}x + (3)6 = (3)9$

$x + 18 = 27$

$x + 18 + ({}^-18) = 27 + ({}^-18)$

$x = 9$

Solve $\frac{1}{5}x - 3 = 4$.

$(5)(\frac{1}{5}x - 3) = 4(5)$

$(5)\frac{1}{5}x - (5)3 = (5)4$

$x - 15 = 20$

$x - 15 + (15) = 20 + 15$

$x = 35$

---

* See Introduction, p. 1.

71. Solve $x - \frac{1}{4} = 7$.

$$(4)(x - \tfrac{1}{4}) = (4)7$$
$$(4)x - (4)\tfrac{1}{4} = (4)7$$
$$4x - 1 = 28$$
$$4x - 1 + (1) = 28 + (1)$$
$$4x = 29$$
$$(\tfrac{1}{4})4x = (\tfrac{1}{4})29$$
$$x = \tfrac{29}{4}$$

72. Solve $2x - 4 = x - \frac{1}{5}$.

$$(5)(2x - 4) = (5)(x - \tfrac{1}{5})$$
$$(5)(2x) - (5)(4) = (5)(x) - (5)(\tfrac{1}{5})$$
$$10x - 20 = 5x - 1$$
$$10x - 20 + (20) = 5x - 1 + (20)$$
$$10x = 5x + 19$$
$$10x + (^-5x) = 5x + (^-5x) + 19$$
$$5x = 19$$
$$(\tfrac{1}{5})5x = (\tfrac{1}{5})19$$
$$x = \tfrac{19}{5}$$

73. If there is more than one fraction, we could multiply in turn by each denominator to eliminate the fractions. It would be much simpler, however, to multiply once by a number that will eliminate all the denominators. The least common denominator is such a number. The least common denominator of the fractions in $\frac{1}{3}x + 2 = 2x - \frac{2}{5}$ is 15 since 15 is the least common multiple of 3 and 5.

$$(15)(\tfrac{1}{3}x + 2) = (15)(2x - \tfrac{2}{5})$$
$$(15)\tfrac{1}{3}x + (15)2 = (15)2x - (15)\tfrac{2}{5}$$
$$5x + 30 = 30x - 6$$

Complete the solution of
$\frac{1}{3}x + 2 = 2x - \frac{2}{5}$.

$$5x + 30 + (^-30) = 30x - 6 + (^-30)$$
$$5x = 30x - 36$$
$$5x + (^-30x) = 30x + (^-30x) - 36$$
$$^-25x = {}^-36$$
$$\left(\frac{^-1}{25}\right){}^-25x = \left(\frac{^-1}{25}\right){}^-36$$
$$x = \tfrac{36}{25}$$

74. As you can see from the frames above the best procedure to use in solving fractional equations is

---

   (a)   Find the least common multiple of the denominators (L.C.D.).

   (b)   Multiply each term in the equation by the L.C.D. to eliminate the fractions.

   (c)   Use the opposites to get all the x terms on one side and all the numbers on the other side of the equals sign.

   (d)   Multiply both sides by the reciprocal of the number in front of x.

---

75. Solve $\frac{2}{9}x - 5 = 7$.

    (a) L.C.D. = 9

    (b) $(9)\frac{2}{9}x - (9)5 = (9)7$

           $2x - 45 = 63$

    (c) $2x - 45 + (45) = 63 + (45)$

           $2x = 108$

    (d) $x = 54$

76. Solve $\frac{3}{5}x - 2 = {}^-7$.

    (a) L.C.D. = 5

    (b) $(5)\frac{3}{5}x - (5)2 = (5)({}^-7)$

           $3x - 10 = {}^-35$

    (c) $3x - 10 + 10 = {}^-35 + 10$

           $3x = {}^-25$

    (d) $(\frac{1}{3})3x = (\frac{1}{3}){}^-25$

           $x = \dfrac{{}^-25}{3}$

77. Solve $\frac{1}{12}x - 3 = \frac{2}{21}x$.

    (a) L.C.D. = 84

    (b) $(84)\frac{1}{12}x - (84)3 = (84)\frac{2}{21}x$

           $7x - 252 = 8x$

    (c) $7x + ({}^-7x) - 252$

           $= 8x + ({}^-7x)$

           ${}^-252 = x$

78. Solve $3x - \frac{1}{7} = \frac{3}{14}$.

    (a) L.C.D. = 14

    (b) $(14)3x - (14)\frac{1}{7} = (14)\frac{3}{14}$

           $42x - 2 = 3$

    (c) $42x - 2 + (2) = 3 + (2)$

           $42x = 5$

    (d) $(\frac{1}{42})42x = (\frac{1}{42})5$

           $x = \frac{5}{42}$

79. To solve $3/x = 9$, we note that the denominator is x. Following the usual procedure we multiply both sides of the equation by x to find $(x)(3/x) = 9(x)$ or $3 = 9x$ (since $3x/x = 3$).

    Solve $3 = 9x$.

    $(\frac{1}{9})3 = (\frac{1}{9})9x$

    $\frac{1}{3} = x$

80. Solve $5/x = 14$.

    $(x)\dfrac{5}{x} = (x)14$

    $5 = 14x$

    $(\frac{1}{14})5 = (\frac{1}{14})14x$

    $\frac{5}{14} = x$

81. Solve $12/x = 6$.

    $(x)\dfrac{12}{x} = (x)6$

    $12 = 6x$

    $(\frac{1}{6})12 = (\frac{1}{6})6x$ or $2 = x$

82. Solve $\frac{3}{7} - (2/x) = 5$.

(a) L.C.D. $= 7x$

(b) $(7x)\left(\dfrac{3}{7} - \dfrac{2}{x}\right) = (7x)5$

$(7x)\dfrac{3}{7} - (7x)\dfrac{2}{x} = (7x)5$

$3x - 14 = 35x$

(c) $3x + (^-3x) - 14$
$= 35x + (^-3x)$
$^-14 = ^-32x$

(d) $(\frac{1}{32})^-14 = (\frac{1}{32})32x$

$\dfrac{^-7}{16} = x$

83. Solve $\frac{6}{11} - (3/x) = {}^-1/2x$.

(a) L.C.D. $= 22x$

(b) $(22x)\dfrac{6}{11} - (22x)\dfrac{3}{x} = (22x)\dfrac{^-1}{2x}$

$12x - 66 = {}^-11$

(c) $12x - 66 + (66)$
$= {}^-11 + (66)$
$12x = 55$
$(\frac{1}{12})12x = (\frac{1}{12})55$
$x = \frac{55}{12}$

Exercise Set 5-3. Work all the problems before checking your answers. Solve each of the following equations for x.

1. $\frac{1}{3}x = {}^-3$

2. $2x = \dfrac{^-3}{7}$

3. $\dfrac{^-5}{x} = 7$

4. $\dfrac{^-7}{3}x = 2$

5. $\frac{2}{3}x - 5 = 7$

6. $\dfrac{^-7}{5}x + 2 = \dfrac{1}{2}$

7. $\frac{1}{15} - \frac{2}{9}x = 4$

8. $\frac{2}{3}x - \frac{4}{9} = \frac{1}{6}x + 2$

9. $\dfrac{3}{x} - 2 = 7$

10. $\dfrac{5}{2x} - \dfrac{3}{7} = \dfrac{1}{14}$

5-4 FORMULAS

Mathematics has provided us with many formulas to help us in solving various problems. To be able to use them fully, however, you must be able to manipulate the formulas. In this section you will learn to work with formulas or literal equations.

84. To solve a literal equation such as $kx = 5$ for a specific letter, we treat each letter other than the one we are solving for as though it were a number. Thus, if we want to solve $kx = 5$ for x, we treat the k as though it were a number, such as 3. To solve $3x = 5$, we multiply each side by $\frac{1}{3}$ to find $(\frac{1}{3})3x = (\frac{1}{3})5$ and $x = \frac{5}{3}$. Solve $kx = 5$ for x.

$$\left(\frac{1}{k}\right)kx = \left(\frac{1}{k}\right)5$$

$$x = \frac{5}{k}$$

85. Solve $xy = k$ for x.

$$\left(\frac{1}{y}\right)xy = \left(\frac{1}{y}\right)k$$

$$x = \frac{k}{y}$$

86. Solve $d = rt$ for t.

$$\left(\frac{1}{r}\right)d = \left(\frac{1}{r}\right)rt$$

$$\frac{d}{r} = t$$

87. In general, to solve a literal equation for a specified variable or letter,

(a) Find the L.C.D. (if there are any fractions).
(b) Multiply each term by the L.C.D. to eliminate the fractions.
(c) Use the opposites to place all terms containing the letter for which you are solving on one side and all other terms on the opposite side.
(d) Multiply each term by the reciprocal of the quantity in front of the letter for which you are solving.

Example: Solve $bx/y - 3 = a$ for x.

(a) L.C.D. $= y$

(b) $(y)\dfrac{bx}{y} - 3(y) = a(y)$

$bx - 3y = ay$

(c) $bx - 3y + (3y) = ay + (3y)$

$bx = ay + 3y$

(d) $\left(\dfrac{1}{b}\right)bx = \left(\dfrac{1}{b}\right)ay + \left(\dfrac{1}{b}\right)3y$

$x = \dfrac{ay}{b} + \dfrac{3y}{b}$

88. A basic law of current flow in an electric circuit is Ohm's law, which states

$$I = \frac{E}{R}$$

where I is the amperage or strength of the current, E is the voltage, and R is the resistance. Derive an equivalent formula that expresses the voltage in terms of the amperage and resistance (i.e., solve for E).

L.C.D. $= R$

$I(R) = \dfrac{E}{R}(R)$

$IR = E$

89. Solve $kx - y = b$ for x.

$x = \dfrac{b}{k} + \dfrac{y}{k}$

90. The efficiency of a machine is expressed as a percentage determined from the ratio of the output to the input. This is expressed by the formula

$$E = \frac{O}{I}$$

Solve this equation for the input, I.

$E(I) = \dfrac{O}{I}(I)$

$EI = O$

$\left(\dfrac{1}{E}\right)EI = \left(\dfrac{1}{E}\right)O$

$I = \dfrac{O}{E}$

Exercise Set 5-4. Work all the problems before checking your answers.

1. Solve $x + y = 9$ for y. _____

2. Solve $3x = k$ for x. _____

3. Solve $kx = y$ for x. _____

4. Solve $2x - y = 7$ for y. _____

5. Solve $d = rt$ for r. _____

6. Solve $k/x = y + t$ for k. _____

7. Solve $3/x + k/x = y$ for x. _____

8. Solve $x/y = m/n$ for y. _____

9. Solve $x - 3y = k$ for y. _____

10. Solve $c = \pi rs/6$ for s. _____

11. Solve $I = E/Z$ for Z. _____

SELF-TEST

1. Simplify each of the following:
    (a) $2x - 7x$ _____
    (b) $^-5x + {}^-3x$ _____
    (c) $^-3(2x)$ _____
    (d) $\dfrac{^-1}{3}(3x)$ _____

2. Solve for x:
    (a) $3x = {}^-9$ _____
    (b) $x - 7 = 2$ _____
    (c) $3x - 5 = 7$ _____
    (d) $2x = {}^-9$ _____
    (e) $4 - 3x = 7$ _____
    (f) $5 + 6x = {}^-2x$ _____
    (g) $^-3x + 2 + 6x = 7$ _____
    (h) $\frac{1}{7}x = {}^-4$ _____
    (i) $\dfrac{2}{x} = {}^-7$ _____
    (j) $\dfrac{^-3}{4}x + 2 = {}^-5$ _____
    (k) $\dfrac{7}{x} - \dfrac{3}{5} = \dfrac{1}{10}$ _____
    (l) $x - y = 6$ _____
    (m) $xy = m$ _____

(n) $\dfrac{k}{x} = r$

_____

(o) $\dfrac{5}{x} + \dfrac{k}{x} = y$

_____

APPLIED PROBLEMS

A. Horsepower, H, is defined as force (in pounds) exerted, F, times the distance (in feet) it is exerted, D, divided by the time (in seconds) it takes, T, times 550, or $H = F \cdot D/550T$. Find the missing values in the table below.

	H	F	D	T
1.	10	200	5	
2.		270	300	4
3.	50	300	5,280	
4.	15	5,200		30
5.	72		5,280	60
6.	$\frac{1}{2}$	50		2

Example: 1.   Replacing H, F, and D by 10, 200, and 5, respectively, in the formula, you find

$$10 = \frac{200 \cdot 5}{550T} = \frac{1,000}{550T}$$

Solving for T, you find

$$5,500T = 1,000 \quad \text{or} \quad T = \frac{1,000}{5,500} = \frac{2}{11} \text{ seconds}$$

B. The S.A.E. formula for horsepower is $H = D \cdot D \cdot N/2.5$, where D is the cylinder bore (in inches) and N is the number of cylinders. Use this formula to find the missing values in the table below.

	H	D	N
1.	28.8	3	
2.		2.75	6
3.	39.2	$3\frac{1}{2}$	
4.	40		4
5.		3.6	8

C. A third formula for determining horsepower is $H = PLAN/396{,}000$, where P is the average pressure in pounds per square inch (p.s.i.), L is the length of the stroke in inches, A is the circular surface area of the piston in square inches, and N is the number of power strokes per minute. Find the missing values in the table below.

	H	P	L	A	N
1.	57.5	110	$2\frac{7}{8}$	8	
2.	94.5	100	$3\frac{1}{4}$		9,600
3.		120	$3\frac{3}{8}$	8.3	5,000
4.	36.9	100		6.4	7,500
5.	82.4		3.52	7.8	9,500

D. Solve the following gear formulas for the specified term.

1. $\text{D.P.} = \dfrac{N}{\text{P.D.}}$ (for P.D.)

2. $\text{C.P.} = 2(\text{T.W.})$ (for T.W.)

3. $A = \dfrac{1}{\text{D.P.}}$ (for D.P.)

4. $\text{O.D.} = \dfrac{N+2}{\text{D.P.}}$ (for N)

5. $\text{T.S.} = \dfrac{\pi}{2(\text{D.P.})}$ (for D.P.)

6. $C = \dfrac{.157}{\text{D.P.}}$ (for D.P.)

7. $\text{D.P.} = \dfrac{2}{\text{W.D.}}$ (for W.D.)

8. $\text{R.D.} = \text{O.D.} - 2(\text{W.D.}) - 2\,C$ (for W.D.)

Key: A—Addendum
    C—Clearance
    C.P.—Circular Pitch
    D.P.—Diametral Pitch
    N—Number of Teeth
    O.D.—Outside Diameter
    P.D.—Pitch Diameter
    R.D.—Root Diameter
    T.S.—Tooth Space
    T.W.—Tooth Width
    W.D.—Working Depth

E. Using the formulas in Problem D, find A, C, O.D., P.D., R.D., and W.D. for a gear with the diametral pitch and number of teeth specified below.
1. D.P. = 12, N = 72   4. D.P. = 14, N = 84
2. D.P. = 10, N = 70   5. D.P. = 8, N = 64
3. D.P. = 12, N = 84

F. A rule of thumb for the tap drill size of American National Threads is $t = D_M - (1/N)$, where t is the tap drill size, D_M is the major diameter, and N is the number of threads per inch. Find the best tap drill size (nearest $\frac{1}{32}$ inch) for the D_M, N below.
1. $D_M = \frac{1}{4}''$, N = 20   4. $D_M = \frac{3}{8}''$, N = 16
2. $D_M = \frac{3}{4}''$, N = 10   5. $D_M = \frac{1}{2}''$, N = 13
3. $D_M = \frac{5}{16}''$, N = 18

G. The time, T, it takes to mill a piece L inches long, if the feed is F inches per minute, can be found using the formula $T = L/F$. Find the missing values in the table below.

	T	L	F
1.		10	$1\frac{1}{4}$
2.	16	12	
3.	14		.95
4.		15	.88
5.	25	23	

H. Simple interest, I, is calculated by multiplying the principle, p, times the rate, r, times the time, t (in years), or $I = prt$. Solve this formula for
1. p
2. r
3. t

I. Using the formula $I = prt$, find the missing variables in the table below:

	I	p	r	t
1.		$6,000	12%	1 year
2.	$500	$10,000	10%	
3.	$300	$5,000		1 year
4.	$250		6%	$\frac{1}{2}$ year
5.		$3,000	$11\frac{1}{2}$%	30 months

J. To find the total resistance R_T of three resistors, R_1, R_2, R_3, in parallel; use the formula

$$\frac{1}{R_T} = \frac{1}{R_1} + \frac{1}{R_2} + \frac{1}{R_3}$$

	R_T	R_1	R_2	R_3
1.		1	2	1
2.	3	6	8	
3.	1		23	2
4.		$2\frac{1}{2}$	3	2
5.	12	15	25	

K. The cutting speed, s, in feet per minute, of a lathe for a piece of stock of diameter d inches rotating at r revolutions per minute is found using the formula $s = \pi dr/12$, where $\pi \simeq 3.1416$. Solve this equation for
1. d
2. r

L. Using $s = \pi dr/12$, find the missing dimensions in the table below:

	s	d	r
1.		$\frac{1}{2}''$	400
2.	75	$1''$	
3.		$\frac{3}{4}''$	600
4.	120		500
5.	100	$1\frac{1}{4}''$	

M. The formula $s = \pi dr/60$ gives the cutting speed of a lathe in centimeters per second of a piece of stock of diameter d centimeters turning at r revolutions per minute.

	s	d	r
1.		2	400
2.	5	3	
3.		$2\frac{1}{2}$	500
4.	10		600
5.	3	$3\frac{1}{4}$	

6

ratio and proportion (part I)

OBJECTIVES

After completing this chapter the student should be able to

1. Express a comparison of two numbers as a ratio (a) reduced to lowest terms and (b) as a number to 1.
2. Determine whether two ratios are equal.
3. Solve a proportion.
4. Solve word problems that require use of a direct proportion to solve.

(1–6) **1.** Write 6 yards to 2 feet as a ratio of a number to 1.

_____

(7–9) **2.** Solve for x.

(a) $\dfrac{\frac{3}{4}}{x} = \dfrac{7}{\frac{1}{3}}$

_____

(b) $\dfrac{2.4}{5} = \dfrac{4}{x}$

_____

(10–16) **3.** If it takes 3 hours to mill 2 pieces, how long will it take to mill 7 pieces?

_____

6-1 RATIOS

Mathematics is often used to describe, compare, and differentiate various objects in the real world. One mathematical tool for deriving a numerical comparison of two objects is called a _ratio_ and will be very useful to you.

1. A **ratio** is a comparison of two numbers. For example, if the length of one wrench is 12 inches and the length of a second wrench is 6 inches, the ratio is the comparison of 12 inches to 6 inches. This can be expressed three ways.
   (a) 12 inches : 6 inches
   (b) 12 inches to 6 inches
   (c) $\dfrac{12 \text{ inches}}{6 \text{ inches}}$ or 12 inches/6 inches

   If the first wrench was 20 inches long and the second wrench was 8 inches long, express the ratio of the first wrench to the second wrench in three ways.

   (a) 20 inches : 8 inches
   (b) 20 inches to 8 inches
   (c) $\dfrac{20 \text{ inches}}{8 \text{ inches}}$
   or 20 inches/8 inches

2. The ratio is usually expressed reduced to lowest terms. The ratio 12 inches/6 inches can be reduced to $\frac{2}{1}$. Reduce 20 inches/8 inches to lowest terms.

   $\frac{5}{2}$

3. In many of the technological fields, ratios are often expressed as a number to 1 unit, such as 3.5:1. For example, the ratio of 7 to 2, written $\frac{7}{2}$, would be expressed as $\frac{3.5}{1}$ or 3.5:1. This form of the ratio can be found by dividing both numerator and denominator by the denominator since then the denominator is 1. (See the example below.)

$$\frac{7 \div 2}{2 \div 2} = \frac{3.5}{1}$$

Change the ratio $\frac{9}{4}$ to one with a denominator of 1 in the method described above.

$$\frac{9 \div 4}{4 \div 4} = \frac{2.25}{1}$$

4. Change 20:8 to a ratio of a number to 1.

$20 \div 8 : 8 \div 8 = 2.5:1$

5. If the units of the two numbers being compared are not the same, but they can be changed to the same units, this should be done. For example, the ratio of 2 feet to 10 inches is not $\frac{2}{10}$ since 2 feet is more than 10 inches. Since 2 feet is equal to 24 inches, the correct ratio is $\frac{24}{10} = \frac{12}{5}$. Write the ratio of 2 yards to 2 feet as a number to 1.

2 yards = 6 feet
6 feet/2 feet = $\frac{3}{1}$ or 3:1

6. Write 1 yard to 8 inches as a ratio of a number to 1.

1 yard = 36 inches
$\frac{36}{8} = \frac{4.5}{1}$

Exercise Set 6-1.   Work all the problems before checking your answers.

1. Find the ratio of $18.00 to 4 hours.

2. Write $\frac{7}{4}$ as a ratio of a number to 1.

3. Write 1 dollar to 2 dimes as a ratio of a number to 1.

6-2 PROPORTIONS

A **proportion** is a statement that two ratios are equal in value. Thus, $\frac{6}{9} = \frac{2}{3}$ and $\frac{5}{7} = \frac{A}{13}$ are proportions. Proportions are very important because they allow us to find an unknown value, if we know that two ratios must be equal and know one ratio but only half of the second ratio. A proportion is a special type of equation and can be solved using the techniques of Chapter 5.

7. Since a proportion such as

$$\frac{16}{22} = \frac{A}{33}$$

is an equation, it is solved in the same way as equations in Chapter 5.

Example: Solve

$$\frac{16}{22} = \frac{A}{33}$$

for A.
L.C.D. = 66

$$(66)\frac{16}{22} = \frac{A}{33}(66)$$

$$48 = 2A$$

$$24 = A$$

Solve the proportion below for A.

$$\frac{A}{24} = \frac{15}{25}$$

L.C.D. = 600

$$(600)\frac{A}{24} = \frac{15}{25}(600)$$

$$25A = 360$$

$$A = 14.4$$

8. Solve the proportion below for A.

$$\frac{3}{A} = \frac{6}{.5}$$

L.C.D. = .5A

$$(.5A)\frac{3}{A} = \frac{6}{.5}(.5A)$$

$$1.5 = 6A$$

$$.25 = A$$

9. **Example:**

$$\frac{5}{A} = \frac{\frac{1}{4}}{\frac{1}{2}}$$

L.C.D. $= \frac{1}{2}A$

$$\left(\frac{1}{2}A\right)\frac{5}{A} = \frac{\frac{1}{4}}{\frac{1}{2}}\left(\frac{1}{2}A\right)$$

$$\frac{5}{2} = \frac{A}{4}$$

$$(8)\frac{5}{2} = \frac{A}{4}(8)$$

$$20 = 2A$$

$$10 = A$$

Solve the proportion below for A.

$$\frac{\frac{1}{3}}{A} = \frac{\frac{5}{6}}{2}$$

L.C.D. $= 2A$

$$(2A)\frac{\frac{1}{3}}{A} = \frac{\frac{5}{6}}{2}(2A)$$

$$\frac{2}{3} = \frac{5A}{6}$$

$$(6)\frac{2}{3} = \frac{5A}{6}(6)$$

$$4 = 5A$$

$$\frac{4}{5} = A$$

Exercise Set 6-2.   Work all the problems before checking your answers. Solve the following proportions for A.

1. $\dfrac{6}{14} = \dfrac{9}{A}$

    _____

2. $\dfrac{A}{2.5} = \dfrac{5}{12.5}$

    _____

3. $\dfrac{\frac{3}{8}}{A} = \dfrac{\frac{3}{4}}{7}$

    _____

6-3 APPLICATIONS

In order to be able to use the powerful tool that a proportion is, you will have to be able to set up an appropriate proportion using the information in the problem. This requires that the values be placed in the correct relationship to each other within the proportion.   In this section only direct proportions will be considered.

---

10. It will be helpful to you in setting up a proportion to consider the units.   Can a ratio with units dollars per pound be equal to a ratio that is in the units pounds per dollar?

No, since the units would have to be the same.

11. For this reason when setting up a proportion attention should be paid to ensuring that the units are the same.   Consider the following problem.

Example: John completed 5 models in 3 hours.   How many of the same type can he expect to finish in the next 5 hours?   The units in this problem are models and hours.

What is the unknown in this problem?

The number of models he can expect to complete in 5 hours.

12. Refer to frame 11. One of the terms in the problem is "5 models." Is this term related to 3 hours or 5 hours?

3 hours, since he can do 5 models in 3 hours.

13. Refer to frames 11 and 12. Thus, one ratio in the proportion could be 5 models/3 hours. Which would be the other ratio— *A* models/5 hours or 5 hours/*A* models? Why?

A models/5 hours since it has the same units as the other ratio.

14. Refer to frame 11. A proportion that can be used to solve this problem is

$$\frac{5 \text{ models}}{3 \text{ hours}} = \frac{A \text{ models}}{5 \text{ hours}}$$

Solve this proportion for *A*.

$8\frac{1}{3}$ models

$$(15)\frac{5}{3} = (15)\frac{A}{5}$$
$$25 = 3A$$
$$\frac{25}{3} = A$$
$$8\frac{1}{3} = A$$

15. If a 20-foot section of a steel girder weighs 560 pounds, how much will a 45-foot section weigh? (Set up this problem as a proportion and solve for the unknown.)

1,260 pounds

$$\frac{20 \text{ feet}}{560 \text{ pounds}} = \frac{45 \text{ feet}}{A \text{ pounds}}$$
$$(560A)\frac{20}{560} = \frac{45}{A}(560A)$$
$$20A = 25,200$$
$$A = 1,260$$

16. Set up a proportion for this problem and solve. A piece of stock tapers (decreases in diameter) $\frac{1}{2}$ inch in 12 inches. What is the length of a piece that has a taper of $\frac{3}{4}$ inch?

18 inches

$$\frac{\frac{1}{2}''}{12''} = \frac{\frac{3}{4}''}{A''}$$
$$(12A)\frac{\frac{1}{2}}{12} = (12A)\frac{\frac{3}{4}}{A}$$
$$\frac{A}{2} = \frac{36}{4}$$
$$(4)\frac{A}{2} = (4)\frac{36}{4}$$
$$2A = 36$$
$$A = 18$$

Exercise Set 6–3. Work all the problems before checking your answers. Set up each problem as a proportion and solve.

1. If it takes you 5 hours to mill 7 pieces, how long will it take you to mill 30 pieces?

2. If the bill for 1,200 feet of wire is $84.00, how much should 700 feet of the same wire cost?

SELF-TEST

1. Find the ratio of 2,500 revolutions to 6 minutes.

2. Write $\frac{15}{6}$ as a ratio of a number to 1.

3. Write 6 feet to 6 inches as a ratio of a number to 1.

4. Solve the following proportions for A:

(a) $\dfrac{7}{39} = \dfrac{A}{26}$

(b) $\dfrac{3.4}{A} = \dfrac{5}{.4}$

(c) $\dfrac{\frac{5}{8}}{\frac{3}{4}} = \dfrac{5}{A}$

5. If 5 feet of round stock weighs 7 pounds, how much will 12 feet weigh?

6. If a piece of stock tapers $\frac{3}{8}$ inch in 10 inches, what is the length of a piece that has a taper of $\frac{3}{4}$ inch?

APPLIED PROBLEMS

A. The ratio of X revolutions to Y minutes is $Z:1$. Find the missing values in the table.

	X	Y	Z
1.	500	6	
2.		$5\frac{1}{2}$	1,800
3.	6,000		1,400
4.	8,000	24	
5.		$2\frac{1}{3}$	450

Example: 1. $\frac{500}{6} = Z/1$, so $6 \times Z = 500$ or $Z = 83\frac{1}{3}$.

B. The rear axle ratio $X:1$ is equal to the ratio of the number of teeth, Y, in the ring gear to the number of teeth, Z, in the pinion.

	X	Y	Z
1.	3	66	
2.		70	25
3.	2.4	64	
4.	2.8		24
5.		62	20

C. In the hydraulic brake system, if the master cylinder piston has an area of W square inches, the wheel cylinder piston has an area of X square inches and a force of Y pounds is exerted on the master cylinder piston, this will result in a force of Z pounds exerted against the brake lining.

	W	X	Y	Z
1.	2	3	600	
2.	1.5	2		800
3.	$1\frac{3}{4}$	$2\frac{3}{4}$	700	
4.		3.1	800	1,100
5.	2	2.8	900	
6.	1.8		750	1,000

Example: 1.   Since a force of 600 pounds is exerted on the 2 square inch surface, and you want to find the force exerted by the 3 square inch surface, the proportion is $\frac{600}{2} = Z/3$. Solving for Z, we find $2Z = 1,800$, $Z = 900$.

D. If the friction loss for W feet of hose is X pounds per square inch, the friction loss for Y feet of the same type of hose is Z pounds per square inch.

	W	X	Y	Z
1.	200	14	550	
2.		28	100	8
3.	650	45	200	
4.	700		300	$20\frac{1}{2}$
5.	250	18		54

E. If three pulleys of diameters X (pulley 1), Y (pulley 2), and Z (pulley 3) are on the same belt, the speed ratio of pulley 2 to pulley 1 is X/Y, the speed ratio of pulley 3 to pulley 2 is Y/Z, and the speed ratio of pulley 3 to pulley 1 is X/Z. Find these three speed ratios for the following size pulleys:
1. $X = 20''$, $Y = 12''$, $Z = 25''$
2. $X = 40$ cm, $Y = 50$ cm, $Z = 60$ cm
3. $X = 38$ cm, $Y = 60$ cm, $Z = 65$ cm
4. $X = 15\frac{1}{2}''$, $Y = 22''$, $Z = 44''$
5. $X = 40.5$ cm, $Y = 70$ cm, $Z = 60.5$ cm

F. On a scale drawing, X inches represents Y inches. If a certain measurement is Z inches on the drawing, it represents W inches.

	W	X	Y	Z
1.	12	$\frac{1}{8}$	1	
2.		$\frac{1}{2}$	3	$\frac{3}{4}$
3.	10	$\frac{1}{4}$		$\frac{1}{2}$
4.	6		2	$\frac{1}{4}$
5.	16	$\frac{1}{12}$	2	

Hint: Use a proportion.

G. On a scale drawing, W mm represents X mm. If a certain measurement on the drawing is Y mm, it represents Z mm.

	W	X	Y	Z
1.	10	40	35	
2.	1	7		42
3.	4		20	90
4.	3	35	$15\frac{1}{2}$	
5.		20	16	83

H. Some concrete is mixed in the ratio 1:2:3 of cement, *X*, to sand, *Y*, to stone, *Z*. Find the missing dimensions in the table below:

	X	Y	Z
1.	50 lb		
2.	100 lb		
3.		250 lb	
4.			500 lb
5.		750 lb	
6.			825 lb

Example: The ratio of cement to sand is 1:2, so $\frac{1}{2} = 50/Y$. Solving for *Y*, we find *Y* = 100 lb. The ratio of cement to stone is 1:3. Thus, $\frac{1}{3} = 50/Z$ and *Z* = 150.

I. The speed ratio of two gears in mesh is equal to the inverse of the ratio of the number of teeth in the two gears. Thus, if gear 1 has *X* teeth and gear 2 has *Y* teeth, the speed ratio of gear 1 to gear 2 is *Y* : *X* = *Z* : 1 expressed as a number to 1. Find the missing values in the table below.

	X	Y	Y : X	Z : 1
1.	40	32		
2.			25 : 32	
3.		45		1.5 : 1
4.	30			.8 : 1
5.	45	25		

J. The compression ratio is a comparison of the amount of space (shaded area in Figure 6-1[a]) in the cylinder when the piston is at the bottom of the stroke with the amount of space (shaded area in Figure 6-1[b]) in the cylinder when it is at the top of the stroke. Find the missing values in the table below if $X:Y = Z:1$ when expressed as a number to 1.

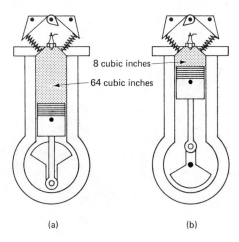

8 cubic inches

64 cubic inches

(a)   (b)

FIGURE 6-1

	X	Y	$Y:X$	$Z:1$
1.	64	8		
2.			58:7.4	
3.		6.4		7.9:1
4.	49.5			8.1:1
5.	75	9		

7

percent

OBJECTIVES

After completing this chapter the student should be able to
1. Define the term *percent*.
2. Change a mixed number to a percent.
3. Change a decimal numeral to a percent.
4. Change a percent to a mixed number or decimal numeral.
5. Solve a given percent problem.
6. Solve a word problem that requires use of a percent in its solution.

(1–13)   **1.**   (a)   Change the following to percents:

(i) $\frac{7}{10}$

(ii)   2.04

(iii)   .0031

_____

_____

_____

(b)   Change $62\frac{1}{2}\%$ to a fraction.

_____

(c)   Change 3.6% to a decimal.

_____

(14–19)   **2.**   Solve the following problem:

7 is 42% of what number?

_____

(20–31)   **3.**   If the sticker price of $3,180 is 6% more than the price of the same car last year, what was the sticker price of last year's model?

_____

7-1   DEFINITIONS

Since many people have an aversion to fractions and decimals, a technique was developed to avoid using them in many situations. Consider, for example, $\frac{1}{4}$, $\frac{3}{25}$, .2, and .17. In each case if we multiply the fraction or decimal by 100, we obtain a whole number (e.g., $\frac{1}{4} \times 100 = 25$). However, 25 is clearly not equal to $\frac{1}{4}$. A symbol must be used to show that we found 25 from $\frac{1}{4}$ by multiplying by 100. This symbol is the percent symbol, $\%$.

---

1.   As stated above we must define a new symbol if we want to have the luxury of avoiding certain fractions. The symbol used, $\%$, is called the _percent symbol_. The term _percent_ tells us exactly what the symbol means. The _per_ means _divided by_ (as in miles _per_ gallon, price _per_ pound) and the _cent_ means _one hundred_ (as in _cent_ury). Thus, **percent** means _divided by one hundred_. Thus, 17% means $17 \div 100$, while 31% means _____.

$31 \div 100$

2.   Since multiplication and division are opposite operations, if we multiply by 100 and divide by 100, the result is equal to the original number. Therefore, $\frac{1}{4} = \frac{1}{4} \times 100 \div 100$. But $\frac{1}{4} \times 100 = 25$, so $\frac{1}{4} = 25 \div 100$. Since $\%$ means "$\div 100$," we can substitute $\%$ for "$\div 100$" to find $\frac{1}{4} = 25\%$. This leads to the following technique for changing a number to a percent.

(a)   Multiply the number by 100 and simplify.
(b)   Place the symbol, $\%$, behind the number.

Change $\frac{1}{5}$ to a percent.

(a) $\frac{1}{5} \times 100 = 20$
(b) $\frac{1}{5} = 20\%$

3. **Example:** Change $\frac{2}{3}$ to a percent.
    (a) $\frac{2}{3} \times 100 = \frac{200}{3} = 66\frac{2}{3}$
    (b) $\frac{2}{3} = 66\frac{2}{3}\%$
Change $\frac{5}{6}$ to a percent.

    (a) $\frac{5}{6} \times 100 = \frac{500}{6} = 83\frac{1}{3}$
    (b) $\frac{5}{6} = 83\frac{1}{3}\%$

4. Change 5 to a percent.

    (a) $5 \times 100 = 500$
    (b) $5 = 500\%$

5. Change $\frac{7}{8}$ to a percent.

    (a) $\frac{7}{8} \times 100 = \frac{700}{8} = 87\frac{1}{2}$
    (b) $\frac{7}{8} = 87\frac{1}{2}\%$

6. **Example:** Change .33 to a percent.
    (a) $.33 \times 100 = 33$
    (b) $.33 = 33\%$
Change .8 to a percent.

    (a) $.8 \times 100 = 80$
    (b) $.8 = 80\%$

7. Change .03 to a percent.

    (a) $.03 \times 100 = 3$
    (b) $.03 = 3\%$

8. Change .017 to a percent.

    (a) $.017 \times 100 = 1.7$
    (b) $.017 = 1.7\%$

9. Change 4.5 to a percent.

    (a) $4.5 \times 100 = 450$
    (b) $4.5 = 450\%$

10. **Example:** Change $2\frac{1}{3}$ to a percent.
    (a) $2\frac{1}{3} \times 100 = \frac{7}{3} \times 100 = \frac{700}{3} = 233\frac{1}{3}$
    (b) $2\frac{1}{3} = 233\frac{1}{3}\%$
Change $1\frac{7}{9}$ to a percent.

    (a) $1\frac{7}{9} \times 100 = \frac{16}{9} \times 100$
         $= \frac{1600}{9} = 177\frac{7}{9}$
    (b) $1\frac{7}{9} = 177\frac{7}{9}\%$

11. Change $4\frac{3}{4}$ to a percent.

    $4\frac{3}{4} = 475\%$

12. The definition of the percent symbol can also be used to change a percent back to a fraction or decimal. Since $\%$ means "$\div 100$," $25\% = 25 \div 100$ or $\frac{25}{100}$. Since $25 \div 100 = .25$, $25\% = .25$ also. Also, since $\frac{25}{100} = \frac{1}{4}$, $25\% = \frac{1}{4}$. Change 30% to a decimal and to a fraction.

    $30\% = 30 \div 100 = .3$
    $30\% = \frac{30}{100} = \frac{3}{10}$

13. **Example:** Change $37\frac{1}{2}\%$ to a fraction.

$$37\frac{1}{2} = \frac{37\frac{1}{2}}{100}$$

$$= \tfrac{75}{2} \div \tfrac{100}{1} = \tfrac{75}{2} \times \tfrac{1}{100}$$

$$= \tfrac{3}{8}$$

Change $12\frac{1}{2}\%$ to a fraction.

$$12\frac{1}{2}\% = \frac{12\frac{1}{2}}{100}$$

$$= \frac{25}{2} \div \frac{100}{1}$$

$$= \frac{25}{2} \times \frac{1}{100}$$

$$= \frac{1}{8}$$

Exercise Set 7-1.   Work all the problems before checking your answers.

1. Change the following fractions or mixed numbers to percents.
    (a) $\frac{3}{5}$
    (b) $\frac{5}{7}$
    (c) $3\frac{1}{6}$

2. Change the following decimals to percents.
    (a) .7
    (b) 1.03
    (c) .004

3. Change the following percents to decimals.
    (a) 40%
    (b) 134%
    (c) 6%
    (d) 3.4%

4. Change the following percents to fractions.
    (a) 16%
    (b) 225%
    (c) $8\frac{1}{3}\%$

7-2 SOLVING PERCENT PROBLEMS

Percent problems are a part of life today. A savings account may pay 5% interest, while the interest on a car loan is 11.98%. A store advertises 20% off on all its furniture. A business offers contractors a 15% discount. In this section you will learn a technique for solving the various types of problems involving percents you may encounter.

---

14. Consider the following problem:
     What is 15% of $34.20? This problem can be restated as "15 is to
     100 as what number is to $34.20," since $15\% = \frac{15}{100}$. This statement

presents the problem as a proportion problem. It can be set up as $\frac{15}{100} = A/\$34.20$. Solve this problem for A.

L.C.D. = 3,420

$$(3,420)\frac{15}{100} = (3,420)\frac{A}{34.20}$$

$$513 = 100A$$

$$5.13 = A$$

(Thus, 15% of \$34.20 is \$5.13.)

15. Refer to frame 14. The number 15 in 15% is called the *rate*, while the number that represents 100% of the total (\$34.20 in this case) is called the *base*. That part of \$34.20 that represents 15% of \$34.20 is called the *percentage*. The percentage was originally called A in frame 14 until it was determined that the percentage was \$5.13. In the problem—"What is 34% of 210?"—state the rate, base, and percentage, if given.

rate: 34
base: 210
percentage: not given

16. Refer to frame 14. In the proportion $\frac{15}{100} = A/\$34.20$, 15 is the rate, A is the percentage, and \$34.20 is the base. If we use r for rate, b for base, and p for percentage and replace the appropriate terms by these abbreviations, we obtain the proportion

$$\frac{r}{100} = \frac{p}{b}$$

This proportion can be used to solve the various types of percent problems.

Example: 16 is 40% of what number?

$$\frac{40}{100} = \frac{16}{b}$$

Solving this proportion for b, we obtain

$$(100b)\frac{40}{100} = (100b)\frac{16}{b}$$

$$40b = 1,600$$

$$b = 40$$

Solve the problem

14 is 25% of what number?

$$\frac{25}{100} = \frac{14}{b}$$

$$(100b)\frac{25}{100} = (100b)\frac{14}{b}$$

$$25b = 1,400$$

$$b = 56$$

17. Use the general proportion given in frame 16 to solve the following problem:

    35 is what percent of 84?

$$\frac{r}{100} = \frac{p}{b}$$

$$\frac{r}{100} = \frac{35}{84}$$

$r = 41\frac{2}{3}$ (Thus, 35 is $41\frac{2}{3}\%$ of 84.)

18. What is 17% of 48?

$$\frac{r}{100} = \frac{p}{b}$$

$$\frac{17}{100} = \frac{p}{48}$$

$p = 8.16$ or $8\frac{4}{25}$

19. 38 is 16% of what number?

$$\frac{16}{100} = \frac{38}{b}$$

$b = 237\frac{1}{2}$

Exercise Set 7-2. Work all the problems before checking your answers.

1. State the proportion to be used in solving percent problems.

2. What is 7.2% of 49?

3. 6 is $37\frac{1}{2}\%$ of what number?

4. 10 is what percent of 45?

7-3 APPLICATIONS

Problems involving percents will not always be stated in the same format. Quite often you must restate the problem in a format that can readily be translated into a proportion and solved. In this section several examples will be worked to demonstrate this idea.

---

20. Consider the following problem: The total bill is $101.41. If there is a 3% discount if you pay in cash, how much would you save by paying for the materials in cash? If this problem is restated in the format "p is $r\%$ of b," it would be easy to solve for the amount of savings by the technique outlined in Section 7-2. The key to doing this is to ask the questions
    (a) "What is the rate?"
    (b) "Of which quantity are we taking a certain percent?"
    The reason for finding the rate first is that it is the most obvious variable. What is the rate in the problem above?

    3 (or 3%)

21. Refer to frame 20. Phrases such as *how much, how many, what amount* and *what percent* refer to the unknown quantity. In the problem stated above, *how much* refers to the unknown amount that will be saved. Will the seller take 3% of this amount or 3% of the $101.41?

3% of $101.41

22. From frame 21 we see that $101.41 is the base, so the unknown is the percentage, p. Restate the problem in the form "p is r% of b."

p is 3% of $101.41, since $r = 3$ and $b = \$101.41$.

23. Solve the problem in frame 20. (Round the answer off to the nearest cent.)

$$\frac{r}{100} = \frac{p}{b}$$

$$\frac{3}{100} = \frac{p}{\$101.41}$$

$3 \cdot 101.41 = 100p$

$304.23 = 100p$

$3.04 = p$

24. State the problem below in the standard form, "p is r% of b" and solve for the unknown. A micrometer is priced at $39.90. If this is 80% of the list price, what is the list price?

$39.90 is 80% of b.

$$\frac{80}{100} = \frac{\$39.90}{b}$$

$80b = \$3990$

$$b = \frac{\$3990}{80} = \$49.88$$

25. Consider the problem in frame 24 again. The price of $39.90 was 20% *less than* the list price of $49.88. However, $39.90 is *not* 20% of $49.88. Consider the similar problem below. All tools are 30% *off* their regular price during the clearance sale. If a circular saw is on sale for $42.70, what was its original price? Notice that $42.70 is *not* 30% of the regular price but 30% off or less than the regular price. *When the given rate is taken off or subtracted from the original amount, the rate desired in the form "p is r% of b" can be found by subtracting the given percent from 100%.* What is the rate, r, in this problem?

100% − 30% = 70%

26. Set up and solve the problem in frame 25.

$42.70 is 70% of b.

$$\frac{70}{100} = \frac{\$42.70}{b}$$

$70 \times b = \$4{,}270$

$b = \$61.$

27. Jim's profit of $39,100 was 10% less than last year's profit. What was last year's profit (to the nearest hundred dollars)?

$$100\% - 10\% = 90\%$$
$39,100 is 90% of b

$$\frac{90}{100} = \frac{\$39,100}{b}$$

$$b = \$43,400$$

28. 4,185 usable parts were produced. If 7% of the parts produced were defective, how many parts were produced altogether? Set up this problem and solve it.

$$100\% - 7\% = 93\%$$
4,185 is 93% of b.

$$\frac{93}{100} = \frac{4,185}{b}$$

$$b = 4,500$$

29. The following problem represents another type of percent problem. A contractor adds 25% onto the wholesale price he pays for materials when itemizing a bill for doing a job. If the total retail cost of the materials on the itemized bill is $835, how much did the materials cost the contractor? In this case the contractor takes 25% of the wholesale price. But $835 is *not* 25% of the wholesale price since it is greater than the wholesale price. In fact, $835 is the wholesale price plus 25% of the wholesale price. Let w be the wholesale price, then 100% of $w = \frac{100}{100} \times w = w$, so we can replace w by 100% of w. Thus, $835 is 100% of w + 25% of w or 125% of w. Therefore, the correct rate to use in solving the problem above is not 25% but

_____.

125%

30. *Whenever a certain percent is "added on" to an original amount, the rate to use in solving the problem is 100% plus the given rate.* Solve the problem in frame 29.

$835 is 125% of b.

$$\frac{125}{100} = \frac{\$835}{b}$$

$$125 \times b = \$83,500$$
$$b = \$668$$

31. The total bill including the 6% sales tax is $31.80. How much did the purchases cost excluding the sales tax? (*Note*: The 6% was "added on" to the purchase price.)

$31.80 is 106% of b.

$$\frac{106}{100} = \frac{\$31.80}{b}$$

$$106b = \$3,180$$
$$b = \$30.$$

Exercise Set 7-3.   Work all the problems before checking your answers.
Solve the problems below for the unknown.

1. If an error of 3% is acceptable and the diameter is supposed to be
2.25″, how much could the diameter vary from this value?

2. If 3 pounds of lime is added for every 20 pounds of cement in
making a mortar, the lime is what percent of the cement used?

3. If you will need 220 pounds of dry mortar in which lime in the
amount of 10% of the weight of the cement used will be added to
the cement, how much cement will you need?

4. How many of the 725 parts produced were usable if 4% of them
were defective?

5. What was the original price, if the price of $255 is 15% less than
the original price?

SELF-TEST

1. Change the following fractions or mixed number to percents:
(a) $\frac{5}{16}$
(b) $\frac{3}{10}$
(c) $2\frac{1}{9}$

2. Change the following decimals to percents:
(a) .035
(b) .3
(c) 3

3. Change the following percents to decimals:
(a) 32%
(b) .02%
(c) 230%
(d) 4%

4. Change the following percents to fractions:
(a) 14%
(b) 350%
(c) $11\frac{3}{5}\%$

5. What is 6.4% of 38?

6. 8 is $14\frac{2}{3}\%$ of what number?

7. 14 is what percent of 11?

8. If the output of an engine is 28 horsepower and the theoretical
horsepower is 40 horsepower, what percent of the theoretical
horsepower is the output?

9. A certain alloy is 78% tin. If a piece of this alloy weighs 15 pounds, what is the weight of the tin in this alloy?

   _____

10. The pitch of a gear was 3% larger than it should have been. If the pitch was 1.8025", what should it have been?

   _____

11. If the pitch of the gear in Problem 10 should have been 2", what was the pitch of the gear?

   _____

APPLIED PROBLEMS

A. The efficiency ($E\%$) of an engine is the ratio of the useful energy output (O) to the energy input (I). Find the missing values in the table below.

	E	O	I
1.	80	4 H.P.	
2.	65		45 H.P.
3.		72 H.P.	95 H.P.
4.	70.2	2.3 H.P.	
5.		$89\frac{1}{2}$ H.P.	120 H.P.

Example: 1. The proportion $E/100 = O/I$ is used. $\frac{80}{100} = 4/I$. Solving for I, we find $80 \times I = 400$ or $I = 5$ horsepower.

B. The thermal efficiency, E_T, of an engine is defined as the ratio of the amount of heat transformed into work, O_H, to the amount of heat produced, I_H. Find the missing values in the table below. (*Note*: BTU is the abbreviation for British Thermal Unit, a unit of heat or energy.)

	E_T	O_H	I_H
1.		43,000 BTU	89,000 BTU
2.	44%		75,000 BTU
3.	$65\frac{1}{2}\%$	48,500 BTU	
4.		61,000 BTU	99,000 BTU
5.	48.7%		68,000 BTU

C. After discounts of first $W\%$ and then $X\%$, an item that had cost Y dollars only cost Z dollars. Find the missing values in the table below.

	W	X	Y	Z
1.	20	10	120	
2.	15	$7\frac{1}{2}$	270	
3.	30		145	91.50
4.		3	75	65.48
5.	10	5		76.95
6.	20	15		71.40

Example: 1. To find the cost that is 20% less than $120, we find 80% of $120 using $\frac{80}{100} = A/120$ and get $96. After taking the 10% discount using $\frac{90}{100} = Z/96$, we find $86.40 for the final cost, Z.

D. Of the X parts produced, Y were defective, so Z percent were defective.

	X	Y	Z
1.	130	6	
2.		14	3
3.	298		5
4.	386	13	
5.		10	$2\frac{1}{2}$

E. The contractor's price of X dollars was Y percent less than the list price of Z dollars.

	X	Y	Z
1.		20	9.95
2.	85	25	
3.	60		72
4.		10	15.50
5.	32.25	$12\frac{1}{2}$	
6.	42.50		65

8

exponential notation

OBJECTIVES

After completing this chapter the student should be able to
1. Define the terms *exponent* and *base*.
2. Write the product of a factor used several times in exponential form.
3. Write an exponential expression as an expanded product.
4. Find the product of two exponential numbers with the same base and signed number exponents.
5. Find the quotient of two exponential numbers with the same base and signed number exponents.
6. Simplify an exponential number raised to a power.
7. Use exponential notation to help find square roots and cube roots.

PRETEST 8 (1–72)

(1–18) **1.** Evaluate 4^3. _____

(19–39) **2.** Simplify the following:

  (a) $\dfrac{x^7}{x^3}$ _____

  (b) $(x^4)^3$ _____

(40–50) **3.** Simplify the following (use positive exponents):
  (a) 5^0 _____
  (b) x^{-3} _____

(51–72) **4.** Find $\sqrt[3]{-64}$ _____

8-1 DEFINITION

You have already seen the operation of multiplication. It is a way of simplifying repeated addition. In this chapter we introduce another notation to simplify repeated multiplication. It is called *exponential notation* and is defined in this section.

---

1. $3 + 3 + 3 + 3 + 3 + 3 + 3 + 3 = 8(3)$
    $6 + 6 + 6 + 6 + 6 + 6 + 6 =$ _____ $7(6)$

2. Which is faster—adding 3 eight times or multiplying 8 by 3?   Multiplying 8(3), once you know the multiplication table.

3. Quite often in working with numbers we are asked to use one number as a factor many times. For example, we may have to simplify $5 \cdot 5 \cdot 5 \cdot 5$. How many times is 5 used as a factor in $5 \cdot 5 \cdot 5 \cdot 5$?   4

4. If we had a way of indicating that 5 was used as a factor 4 times, we could save time and eventually, as with multiplication, we may memorize some of the simpler problems and not have to go through the long process of multiplication. The notation we use to indicate that 5 is used as a factor 4 times is 5^4. The numeral 4 is called the *exponent* and the 5 is called the *base*. $4 \cdot 4 \cdot 4$ is indicated as 4^3 since 4 is the factor and it is used 3 times. How would you indicate $3 \cdot 3 \cdot 3 \cdot 3 \cdot 3$ in exponential notation? 3^5

5. How often is 7 used as a factor in $7 \cdot 7 \cdot 7 \cdot 7$?   4 times

6. Write $7 \cdot 7 \cdot 7 \cdot 7$ in exponential notation. 7^4

152

7. Write $2 \cdot 2 \cdot 2 \cdot 2 \cdot 2 \cdot 2$ in exponential notation. 2^6

8. Write $6 \cdot 6 \cdot 6 \cdot 6 \cdot 6$ in exponential notation. 6^5

9. Write $x \cdot x \cdot x \cdot x$ in exponential notation. x^4

10. Write $r \cdot r \cdot r$ in exponential notation. r^3

11. Write $d \cdot d$ in exponential notation. d^2

12. $5^3 = 5 \cdot 5 \cdot 5$. Write 6^4 in this expanded form. $6^4 = 6 \cdot 6 \cdot 6 \cdot 6$

13. Write 7^5 in expanded form. $7 \cdot 7 \cdot 7 \cdot 7 \cdot 7$

14. Write r^4 in expanded form. $r \cdot r \cdot r \cdot r$

15. Example: To evaluate 5^3 means to find the value of the expression.
    Thus $5^3 = 5 \cdot 5 \cdot 5 = 125$.
    Evaluate 4^3. $64 \ (4 \cdot 4 \cdot 4 = 64)$

16. Evaluate 3^4. 81

17. Evaluate 5^1. $5^1 = 5$ since there is only one factor of 5.

18. Evaluate 12^3. $1,728$

Exercise Set 8-1.   Work all the problems before checking your answers.

1. Write each of the following in exponential notation.
   (a) $2 \cdot 2 \cdot 2 \cdot 2$ _____
   (b) $5 \cdot 5 \cdot 5 \cdot 5 \cdot 5$ _____
   (c) $k \cdot k \cdot k$ _____

2. Write each of the following in expanded form.
   (a) 5^4 _____
   (b) 7^8 _____
   (c) x^5 _____

3. Evaluate each of the following:
   (a) 2^2 _____
   (b) 5^4 _____
   (c) 7^1 _____
   (d) 6^3 _____

8-2 LAWS OF EXPONENTS

In this section we shall develop some rules for operating with exponential numbers. All the rules follow logically from the definition of an exponent, however, so you can always simplify an expression involving exponents even if you forget the rules.

---

19. $2^3 \cdot 2^2 = (2 \cdot 2 \cdot 2) \cdot (2 \cdot 2)$ by the definition of an exponent. But
     $2 \cdot 2 \cdot 2 \cdot 2 \cdot 2 = $ _____ in exponential notation. 2^5

20. $3^4 \cdot 3^2 = (3 \cdot 3 \cdot 3 \cdot 3) \cdot (3 \cdot 3) = $ _____ 3^6

21. $5^2 \cdot 5^5 = $ $(5 \cdot 5) \cdot (5 \cdot 5 \cdot 5 \cdot 5 \cdot 5) = 5^7$

22. $7^2 \cdot 7^6 = $ $(7 \cdot 7) \cdot (7 \cdot 7 \cdot 7 \cdot 7 \cdot 7 \cdot 7) = 7^8$

23. Have you noticed any pattern in the problems above? They are
     listed below.
     $2^3 \cdot 2^2 = 2^5$
     $3^4 \cdot 3^2 = 3^6$
     $5^2 \cdot 5^5 = 5^7$
     $7^2 \cdot 7^6 = 7^8$
     In each case we used the same base and _____ the exponents.   added
     <small>(added, multiplied)</small>

24. The rule is the following:

     $$x^a \cdot x^b = x^{a+b}$$

     Thus,

     $$5^3 \cdot 5^6 = 5^{3+6} = 5^9$$

     Simplify $8^3 \cdot 8^7$. $8^{3+7} = 8^{10}$

25. Simplify $4^5 \cdot 4^6$. $4^{5+6} = 4^{11}$

26. The reason we add the exponents is that they tell us the number of
     factors of the number. Thus in $4^9 \cdot 4^5$ we have the product of 9
     factors of 4, times the product of 5 factors of 4, or altogether we have
     the product of 14 factors of 4.
     Simplify $9^3 \cdot 9^2$. $9^{3+2} = 9^5$

27. To simplify $\frac{30}{18}$ we write each number in factored form and divide
     through by common factors.
     Thus,

     $$\frac{30}{18} = \frac{2 \cdot 3 \cdot 5 \div 2 \cdot 3}{2 \cdot 3 \cdot 3 \div 2 \cdot 3} = \frac{5}{3}.$$

$6^5/6^2$ can be simplified in the same way.

$$\frac{6^5}{6^2} = \frac{6 \cdot 6 \cdot 6 \cdot 6 \cdot 6 \div 6 \cdot 6}{6 \cdot 6 \div 6 \cdot 6} = \underline{\hspace{2cm}}$$

$6 \cdot 6 \cdot 6$ or 6^3

28. Refer to frame 27. In simplifying a division problem involving exponential numbers with the same base we $\underline{\hspace{2cm}}$ the $\underset{\text{(subtract/divide)}}{}$ exponents.

subtract

29. The rule is

$$\frac{x^a}{x^b} = x^{a-b}$$

Thus,

$$\frac{5^7}{5^5} = 5^{7-5} = 5^2$$

Similarly,

$$\frac{3^7}{3^2} = 3^{7-2} = \underline{\hspace{2cm}}$$

3^5

30. $\dfrac{11^4}{11^2} = 11^{4-2} = \underline{\hspace{2cm}}$

11^2

31. $x^2 \cdot x^2 = x^{2+2} = x^4$
    $x^2 \cdot x^2 \cdot x^2 = x^{2+2+2} = x^6$
    $x^2 \cdot x^2 \cdot x^2 \cdot x^2 = x^{2+2+2+2} = x^8$
    $x^2 \cdot x^2 \cdot x^2 \cdot x^2 \cdot x^2 =$

$x^{2+2+2+2+2} = x^{10}$

32. $x^2 \cdot x^2 = (x^2)^2$, since x^2 is used as a factor 2 times.
    $x^2 \cdot x^2 \cdot x^2 = (x^2)\underline{\hspace{1cm}}$, since x^2 is used as a factor 3 times.

3

33. $x^2 \cdot x^2 \cdot x^2 \cdot x^2 = \underline{\hspace{2cm}}$

$(x^2)^4$

34. $x^2 \cdot x^2 \cdot x^2 \cdot x^2 \cdot x^2 = \underline{\hspace{2cm}}$

$(x^2)^5$

35. Refer to frame 31.
    $(x^2)^2 = x^{2+2} = x^4$
    But $2 + 2 = 2 \cdot 2$, so $(x^2)^2 = x^{2 \cdot 2}$
    $(x^2)^3 = x^{2+2+2}$, but $2 + 2 + 2 = 2 \cdot 3$,
    so $(x^2)^3 = x$.

$2 \cdot 3$

36. $(x^2)^4 = x^{2+2+2+2} = x^{2 \cdot} \underline{\hspace{0.5cm}}$

4

37. $(x^2)^5 =$ _____ $x^{2 \cdot 5}$

38.

> When using an exponential number as a factor several times, the
> expression can be simplified by multiplying the exponents. That is,
> $$(x^a)^b = x^{a \cdot b}$$

For example, $(3^2)^4 = 3^{2 \cdot 4} = 3^8$

$(2^3)^4 =$ _____ $2^{3 \cdot 4} = 2^{12}$

39. Simplify $(5^4)^3$. $5^{4 \cdot 3} = 5^{12}$

> The laws for dealing with exponents are as follows:
> (a) $x^a \cdot x^b = x^{a+b}$
> (b) $\dfrac{x^a}{x^b} = x^{a-b}$
> (c) $(x^a)^b = x^{a \cdot b}$

Exercise Set 8-2.   Work all the problems before checking your answers.
1. Simplify $3^5 \cdot 3^4$.   _____
2. Simplify $5^7 \cdot 5^4$.   _____
3. Simplify $7^9/7^3$.   _____
4. Simplify x^5/x.   _____
5. Simplify $x^2 \cdot x^3$.   _____
6. Simplify $(5^2)^5$.   _____
7. Simplify $(x^3)^5$.   _____

8-3   SIGNED NUMBER EXPONENTS

There is no reason why numbers such as $\dfrac{1}{5 \cdot 5 \cdot 5}$ cannot be written in exponential

notation. In this section you will see how the introduction of negative exponents
simplifies work with fractional expressions involving exponents. Negative
exponents will be very important in understanding scientific notation and in
working with logarithms.

40. From the laws of exponents that we developed in the last section we
know that $5^3/5^2 = 5^{3-2} = 5^1$ or 5 and $7^9/7^4 = 7^{9-4} = 7^5$.
Simplify $6^5/6^5$. $6^{5-5} = 6^0$

41. We have not talked about an exponent of 0 up to this time. What would be a reasonable definition for 6^0? Since $6^5/6^5 = 1$ (a number divided by itself is always 1, except for 0/0), 6^0 must equal 1. $5^3/5^3 = 1$, but $5^3/5^3 = 5^{3-3} = 5^0$ also.
Thus, 5^0 must equal _____ since $5^0 = 5^3/5^3$.

         1

42. Since it seems that in each case a number with an exponent of 0 is 1, we define x^0 to be 1 for all numbers x. Thus, $5^0 = 1$.
$13^0 =$ _____

         1

43. $8^0 =$ _____

         1

44. From arithmetic we know that

$$\frac{4}{28} = \frac{2 \cdot 2 \div 2 \cdot 2}{2 \cdot 2 \cdot 7 \div 2 \cdot 2} = \frac{1}{7}$$

In the same way,

$$\frac{2^2}{2^5} = \frac{2 \cdot 2 \div 2 \cdot 2}{2 \cdot 2 \cdot 2 \cdot 2 \cdot 2 \div 2 \cdot 2} = \frac{1}{2 \cdot 2 \cdot 2} = \frac{1}{2^3}$$

Thus,

$$\frac{3^3}{3^5} = \frac{3 \cdot 3 \cdot 3 \div 3 \cdot 3 \cdot 3}{3 \cdot 3 \cdot 3 \cdot 3 \cdot 3 \div 3 \cdot 3 \cdot 3} = \underline{\hspace{3cm}}$$

         $\dfrac{1}{3 \cdot 3} = \dfrac{1}{3^2}$

45. The second law of exponents tells us that $x^a/x^b = x^{a-b}$. Thus

$$\frac{2^2}{2^5} = 2^{2-5} = 2^{-3}(2 - 5 = 2 + {}^-5 = {}^-3)$$

But in frame 44 we say that $2^2/2^5 = 1/2^3$. Thus $2^{-3} = 1/2^3$.

$$\frac{3^3}{3^5} = 3^{3-5} = \underline{\hspace{2cm}}$$

         3^{-2}

46. In frame 44 we saw that $3^3/3^5 = 1/3^2$. But in frame 45 we saw that $3^3/3^5 = 3^{-2}$. Therefore, we must have that $3^{-2} =$ _____.

         $1/3^2$

47. Because of the pattern established in frames 44–46, we define negative exponents in the following way:

$$x^{-a} = \frac{1}{x^a}$$

Thus, $5^{-2} = 1/5^2$. In the same way, $3^{-4} =$ _____.

         $\dfrac{1}{3^4}$

48. Since $1/3^4$ is the reciprocal of 3^4 and $3^{-4} = 1/3^4$, we see that 3^{-4} is the reciprocal of 3^4. In other words, *the negative in front of an exponent tells us to take the reciprocal of the exponential number without the negative sign.* For example, the negative in $7^{\ominus 2}$ means *take the reciprocal of 7^2.* Thus, $7^{-2} =$ the reciprocal of $7^2 = $ _____.

$\dfrac{1}{7^2}$

49. $6^{-4} = $ _____

$\dfrac{1}{6^4}$

50. $7^{-5} = $ _____

$\dfrac{1}{7^5}$

Exercise Set 8-3.   Work all the problems before checking your answers.

1. $5^0 = $ _____   _____

2. $x^{-5} = $ _____   _____

3. $x^0 = $ _____   _____

4. $7^{-3} = $ _____   _____

5. $\dfrac{x^2}{x^7} = $ _____   _____

6. $\dfrac{3^4}{3^7} = $ _____   _____

8.4 SQUARE ROOTS AND CUBE ROOTS

In mathematics every operation has an inverse or opposite operation. The opposite of addition is subtraction. The opposite of multiplication is division. The opposite of the square of a number is the square root. In this section we shall look at the operations *square root* and *cube root*.

51. The formula for the area of a square with sides of length s is $A = s^2$. Thus, if the side of a square is 3 inches, the area is 3^2 or 9 square inches. The area of the square in Fig. 8-1 is 16. What is the length of one side?

4

FIGURE 8-1

52. Notice that $4^2 = 16$. 4 is called a *square root* of 16 since when squared it is equal to 16. A square root of 25 is 5 since 5 squared is equal to 25. What is a square root of 4?

    2, since $2^2 = 4$.

53. By squaring a side we can find the area of a square. By taking the square root of the area we can find the length of a side. If the area of a square is 36, what is the length of one side?

    6, since $6^2 = 36$.

54. The symbol used to indicate the square root operation is $\sqrt{}$. $\sqrt{25}$ is read as *the square root of* 25.
    Find $\sqrt{49}$.

    7, since $7^2 = 49$.

55. Find $\sqrt{9}$.

    3, since $3^2 = 9$.

56. Find $\sqrt{64}$.

    8, since $8^2 = 64$.

57. $(^-3)^2 = $ _____

    9

58. From frame 57, $(^-3)^2 = 9$. Thus, a square root of 9 is $^-3$. In fact, there are two square roots for every number—a positive and a negative one. The square roots of 9 are 3 and $^-3$. What are the square roots of 36?

    6, $^-6$

59. What are the square roots of 81?

    9, $^-9$

60. What are the square roots of 121?

    11, $^-11$

61. What are the square roots of 144?

    12, $^-12$

62. To indicate which square root we are talking about, we write $^-\sqrt{25}$ for $^-5$ and simply $\sqrt{25}$ for 5, the positive value. Find $^-\sqrt{49}$.

    $^-7$

63. Find $\sqrt{100}$.

    10

64. Find $^-\sqrt{81}$.

    $^-9$

65. There are many numbers for which we cannot find an exact square root. For example, $\sqrt{7}$ cannot be found exactly. Such numbers are called *irrational*. Can $\sqrt{8}$ be found exactly?*

    no

---

* A square root table or a calculator is used to find approximate decimal values for irrational numbers.

66. $$x^3 = x \cdot x \cdot x$$

    The opposite operation of the cube of a number is the **cube root**. The cube root of 8 is 2 since $2^3 = 8$. The cube root of 27 is _____ since $3^3 = 27$.

    3

67. The cube root of $^-27$ is $^-3$ since $(^-3)^3 = {}^-27$. Find the cube root of $^-8$.

    $^-2$ since $(^-2)^3 = {}^-8$.

68. There is only one real cube root for every number. The symbol to indicate the cube root is $\sqrt[3]{}$, where the ^{3} indicates that we are taking the third or cube root. In the same way we can represent the fourth root as $\sqrt[4]{}$. How would you represent the fifth root?

    $\sqrt[5]{}$

69. Find $\sqrt[3]{64}$.

    4 since $4^3 = 64$.

70. Find $\sqrt[3]{^-64}$.

    $^-4$ since $(^-4)^3 = {}^-64$.

71. Find $\sqrt[3]{125}$.

    5 since $(5)^3 = 125$.

72. Find $\sqrt[3]{1}$.

    1 since $1^3 = 1$.

Exercise Set 8-4.   Work all the problems before checking your answers.

1. Find $\sqrt{25}$. _____
2. Find $\sqrt{121}$. _____
3. Find $\sqrt[\ ]{81}$. _____
4. Which of the following are irrational?
   (a) $\sqrt{10}$ _____
   (b) $\sqrt[\ ]{64}$ _____
5. Find $\sqrt[3]{^-64}$. _____
6. Find $\sqrt[3]{125}$. _____

SELF-TEST

1. Write $4 \cdot 4 \cdot 4 \cdot 4 \cdot 4 \cdot 4$ in exponential notation. _____
2. Write $x \cdot x \cdot x \cdot x \cdot x$ in exponential notation. _____
3. Write x^7 in expanded notation. _____
4. Simplify each of the following:
   (a) 3^4 _____
   (b) $2^7 \cdot 2^5$ _____

(c) $\dfrac{x^6}{x^2}$ _____

(d) $(6^2)^7$ _____

5. Simplify each of the following:
   (a) 7^0 _____

   (b) $\dfrac{a^3}{a^9}$ _____

6. Write x^{-4} with a positive exponent. _____

7. Simplify each of the following:
   (a) $\sqrt{81}$ _____
   (b) $^-\sqrt{49}$ _____

8. Is $\sqrt{30}$ irrational? _____

APPLIED PROBLEMS

A. A formula for determining one large diameter hose that will carry the same amount of water as two other hoses of a smaller diameter is $D = \sqrt{d_1^2 + d_2^2}$, where D is the diameter of the large hose and d_1 and d_2 are the diameters of the two smaller hoses. Use this formula to find the equivalent large diameter for the following smaller hoses.

	d_1	d_2	D
1.	3″	4″	
2.	2″	$1\frac{1}{2}$″	
3.	2.4″	1″	

B. The velocity, V feet per second, of an object dropped from a height of H feet is found using $V = 8\sqrt{h}$. Find V for the following heights.

	h	V
1.	36′	
2.	100′	
3.	169′	
4.	400′	
5.	49′	

C. The relative friction loss factor from one size hose to another is
$F = d_2^5/d_1^5$, where d_1 and d_2 are the diameters of the first and second
hoses, respectively. Find the relative friction loss, F in each of the
following cases.

	d_1	d_2	F
1.	$2\frac{1}{2}''$	$3''$	
2.	$3''$	$2''$	
3.	$5''$	$3''$	
4.	$2\frac{1}{2}''$	$1\frac{1}{2}''$	
5.	$1''$	$2\frac{1}{2}''$	

Example: 1. $F = \dfrac{d_2^5}{d_1^5} = \dfrac{3^5}{(2\frac{1}{2})^5} = \dfrac{3 \cdot 3 \cdot 3 \cdot 3 \cdot 3}{2\frac{1}{2} \cdot 2\frac{1}{2} \cdot 2\frac{1}{2} \cdot 2\frac{1}{2} \cdot 2\frac{1}{2}} = \dfrac{243}{\frac{5}{2} \cdot \frac{5}{2} \cdot \frac{5}{2} \cdot \frac{5}{2} \cdot \frac{5}{2}}$

$= \dfrac{243}{3,125/32} = \dfrac{243}{1} \cdot \dfrac{32}{3,125} \approx 2.5$

D. How many square or cubic inches are there in the following units?
1. square foot
2. cubic foot
3. square yard
4. cubic yard

Example: Since there are 3 feet in a yard, there are $(3)^3$ or 27 cubic feet
in a cubic yard.

E. An interesting function that can be used to find the cube root, fourth
root, etc., when your calculator will not do this is the *logarithm*. The log
(abbreviation for logarithm) of a number, n, to a certain base, b, is defined
to be the exponent, x, such that $b^x = n$. For example, $\log_2 8$ (read *log to
the base 2 of 8*) is 3, since $2^3 = 8$. In the same way, $\log_5 \frac{1}{25} = {}^-2$ since
$5^{-2} = \frac{1}{25}$. Find the logs to the indicated base of the following:
1. $\log_{10} 1,000$
2. $\log_2 32$
3. $\log_{10} \frac{1}{10}$
4. $\log_5 1$
5. $\log_3 27$
6. $\log_3 \frac{1}{27}$
7. $\log_{10} 10^{-4}$
8. $\log_4 \frac{1}{16}$

F. The laws of exponents are

1. $a^n \cdot a^m = a^{n+m}$

2. $\dfrac{a^n}{a^m} = a^{n-m}$

3. $(a^n)^m = a^{nm}$

Notice that $\log_{10}(10^3 \cdot 10^2) = \log_{10} 10^{3+2} = 3 + 2$, while $\log_{10} 10^3 = 3$ and $\log_{10} 10^2 = 2$. Thus, $\log_{10}(10^3 \cdot 10^2) = \log_{10} 10^3 + \log_{10} 10^2$.

In fact, this is true in all cases. Using the other laws of exponents we can show that logarithms have the following properties:

1. $\log_b (x \cdot y) = \log_b x + \log_b y$

2. $\log_b \left(\dfrac{x}{y} \right) = \log_b x - \log_b y$

3. $\log_b (x^n) = n \log_b x$

Use these laws to simplify the following:

1. $\log_{10} \left(\dfrac{120}{.105} \right)$ _____

2. $\log_2 \left(\dfrac{312}{15} \right)$ _____

3. $\log_{10} (16 \cdot 102)$ _____

4. $\log_{10} (13^5)$ _____

5. $\log_3 (5^{1/3})$ _____

6. $\log_{10} (\sqrt[3]{6})$ $(Hint: \sqrt[3]{6} = 6^{1/3})$ _____

Example: 1. $\log_{10} \left(\dfrac{120}{.105} \right) = \log_{10} 120 - \log_{10} .105$

G. You can find the area of a triangle if you know the lengths of the three sides, a, b, and c, using the formula $A = \sqrt{s(s-a)(s-b)(s-c)}$ where $s = \frac{1}{2}(a + b + c)$. Find the area of the triangles with sides as shown below. Leave your answer as a square root, if it is not a perfect square.

	a	b	c	s	A
1.	10	12	14		
2.	2	5	6		
3.	5.3	6	10		
4.	15	25	30		
5.	109.3	100.2	98.3		

H. Using $D = \sqrt{d_1^2 + d_2^2 + d_3^2}$, find a large diameter, D, such that the area of the circle of diameter, D, is the same as the sum of the areas of the circles with diameters $d_1, d_2,$ and d_3.

	d_1	d_2	d_3	D
1.	1″	1″	2″	
2.	.5″	.5″	.75″	
3.	1.5″	1″	2″	
4.	.625″	.5″	1.5″	
5.	.75″	3″	2.5″	

I. The area of a circle can be found using $A = (3.14)(D^2)/4$, where D is the diameter of the circle. Find the area for the following diameters.

	D	A
1.	10.3 ft	
2.	45.7 cm	
3.	210 mm	
4.	8.4 in	
5.	.76 yd	

J. The volume of a sphere of diameter, D, can be found using $V = (3.14)(D^3)/6$. Find the volume for the diameters given in Problem I.

9

the metric system

OBJECTIVES

After completing this chapter the student should be able to
1. State the meaning of the various prefixes used in the metric system.
2. Convert from one metric unit to another with linear, area, and volume units.
3. Convert from a metric unit to an English unit with linear, area, and volume units.
4. Convert from an English unit to a metric unit with linear, area, and volume units.
5. Estimate the metric length, area, etc., of familiar objects.

Solve the following problems using the metric table in the back of the text.

(1-14)   **1.**   950 mm = _____ km

(15-29)   **2.**   900 dm^2 = _____ a

(30-46)   **3.**   (a) 15 kℓ = _____ cℓ

              (b) 61 mm^3 = _____ cm^3

(47-72)   **4.**   (a) 33 in.^{3} = _____ mm^3

              (b) 190 dm^2 = _____ yd^2

(73-91)   **5.**   If the speed limit is 35 miles per hour, approximately how many kilometers per hour does that equal?

9-1   LINEAR METRIC UNITS

The metric system was set up to simplify all the complicated conversions required by the English system. One needs only consider the English units inch, foot, yard, and mile to see that the conversions from one unit to another is not immediate. The metric system avoids this problem since all conversions can be accomplished by simply moving the decimal point. Linear measure in the metric system is based on a constant, the meter. The meter was originally defined as one ten-millionth of the distance of a line from the equator to the north pole through Paris. The metric system is the system of measure in most countries of the world and will be adopted in the United States in the next decade. In this section we look at units of linear measure in the metric system.

---

1.   The meter, which is approximately equal to the yard, is the basic linear unit in the metric system. All other units of linear measure are multiples of 10 of the meter. Prefixes are used to indicate the relation of other units. The table below outlines the prefixes generally used.

Prefix	Meaning
kilo	1,000 or 10^3
hecto	100 or 10^2
deka	10 or 10^1
no prefix	1 or 10^0
deci	1/10 or 10^{-1}
centi	1/100 or 10^{-2}
milli	1/1,000 or 10^{-3}
micro	1/1,000,000 or 10^{-6}

Thus, a *kilo*meter is 1,000 meters = 10^3 meters, while a *deka*meter is _____ meters.

$10 = 10^1$

2. A decimeter is 10^{-1} meter. A centimeter is _____ meter.

10^{-2}

3. A millimeter is _____ meter.

10^{-3}

4. A hectometer is _____ meters.

10^2

5. Abbreviations are often used for the units of linear measure. The abbreviations we shall use are shown in the table below.

Unit	Abbreviation
kilometer	km
hectometer	hm
dekameter	dkm
meter	m
decimeter	dm
centimeter	cm
millimeter	mm
micron	μ

Thus, 5 km means 5 _____.

kilometers

6. 2 mm means 2 _____.

millimeters

7. 11 m means 11 _____.

meters

8. .4 cm means .4 _____.

centimeter

9. 50 μ means 50 _____.

microns

10. To change from one metric unit to another metric unit it will help to use the exponents of 10 indicated by the prefix. For example, to change from kilometers to dekameters you must change from a unit equal to 10^3 meters to a unit equal to _____ meters.

10^1

11. Since a kilometer is 1,000 meters and a dekameter is 10 meters, the number of dekameters in a kilometer is

$$\frac{1{,}000 \text{ meters}}{10 \text{ meters}} = \frac{10^3}{10^1} = 10^{3-1} = 10^2$$

Thus, 1 km = $1 \cdot 10^2$ dkm = 100 dkm; 3.4 km = $3.4 \cdot 10^2$ dkm = 340 dkm; and .74 km = _____ dkm.

$.74 \cdot 10^2$ dkm = 74 dkm

12. Notice in frame 11 that when multiplying by 10^2, the decimal point is moved two places to the right. The exponent of 2 tells us to move the decimal point to the *right* because it is *positive*. Thus, if the exponent is $^-3$, the decimal point is moved 3 places to the *left*; if the exponent of 10 is 4, the decimal point is moved 4 places to the

_____.

right

13. To change from one metric unit to another, you must know how many of the units you are changing to there are in one of the original units. To find this number, you must divide the first unit by the second. Thus, to change hectometers to kilometers, you must divide one hectometer by one kilometer. Since 1 hm $= 10^2$ m and 1 km $= 10^3$ m,

$$\frac{1 \text{ hm}}{1 \text{ km}} = \frac{10^2 \text{ m}}{10^3 \text{ m}} = 10^{2-3} = 10^{2+{}^-3} = 10^{-1}$$

Therefore, 1 hm $= 10^{-1}$ km and 2 hm $= 2 \cdot 10^{-1}$ km $= .2$ km. The illustration above leads to a shortcut method for changing from one metric unit to another.

(a) Write each unit in terms of the meter.
(b) Divide the given metric unit by the one to which you are changing.
(c) Multiply the original number by the quotient found in Step (b) by moving the decimal point the appropriate number of places.

Example: Change 41.6 hm to centimeters
  (a) hm $= 10^2$ m, cm $= 10^{-2}$ m

  (b) $\dfrac{\text{hm}}{\text{cm}} = \dfrac{10^2 \text{ m}}{10^{-2} \text{ m}} = 10^{2-{}^-2} = 10^{2+2} = 10^4$

  (c) 41.6 hm $= 41.6 \cdot 10^4 = 416{,}000.$ cm

Change .03 cm to millimeter

(a) cm $= 10^{-2}$ m; mm $= 10^{-3}$ m

(b) $\dfrac{10^{-2} \text{ m}}{10^{-3} \text{ m}} = 10^{-2-{}^-3}$

  $= 10^{-2+3} = 10^1$

(c) .03 cm $= .03 \cdot 10^1$ mm

  $= .3$ mm

14. .35 hm = _____ mm

(a) hm $= 10^2$ m; mm $= 10^{-3}$ m

(b) $\dfrac{10^2 \text{ m}}{10^{-3} \text{ m}} = 10^{2-{}^-3} = 10^5$

(c) .35 hm $= .35 \cdot 10^5$

  $= 35{,}000$ mm

Exercise Set 9-1. Work all the problems before checking your answers.

1. Define the meaning of each prefix below.
    hecto—100 or 10^2
  (a) kilo— _____
  (b) milli— _____
  (c) micro— _____
  (d) centi— _____

_____
_____
_____
_____

2. Give the abbreviation for each of the following.
   (a) kilometer— _____
   (b) meter— _____
   (c) centimeter— _____
   (d) millimeter— _____

3. Change to the new units in each case.
   (a) 10 dekameters = _____ decimeters
   (b) 5 cm = _____ mm
   (c) 8 m = _____ km
   (d) 9 mm = _____ km

9-2 AREA METRIC UNITS

For area measurements using the metric system, the square meter is the basic unit.
In this section we shall look at metric area measure conversions.

15. The square meter, a square of side 1 meter, is the basic unit of area
    measurement in the metric system and other units are derived from it.
    The unit for land measure is called the *are*. An **are** is a square
    dekameter; that is, it is equivalent to the area of a square of side
    10 meters. Since the area of such a square is 10 meters times 10 meters
    or 10^2 square meters, an are is 10^2 square meters. Since a decimeter
    is 10^{-1} meter, a square decimeter is 10^{-1} meter times
    10^{-1} meter $= (10^{-1})^2$ square meter = _____ square meter.

    10^{-2}

16. Since a millimeter is 10^{-3} meter, a square millimeter is $(10^{-3})^2$
    square meter or 10^{-6} square meter. Similarly, a square centimeter
    = _____ square meter since a centimeter is
    10^{-2} meter.

    $(10^{-2})^2 = 10^{-4}$

17. A square kilometer is _____ square meters since
    one kilometer is 10^3 meters.

    $(10^3)^2 = 10^6$

18. Hecto means 10^2, so a hectare is _____ ares.

    10^2

19. Since an are is 10^2 square meters, a hectare is _____
    square meters.

    $10^2 \cdot 10^2 = 10^4$

20. To abbreviate square meter, square centimeter, square millimeter,
    etc., the previous abbreviations for linear measure are used with an
    exponent of 2 added to indicate that we are talking about area or
    square measurements. Thus, square centimeter is abbreviated cm^2.
    What is the abbreviation for square millimeter?

    mm^2

21. What is the abbreviation for square kilometer?

km^2

22. What is the abbreviation for square hectometer?

hm^2

23. m^2 is the abbreviation for _____ _____.

square meter

24. What is the abbreviation for square decimeter?

dm^2

25. The abbreviations for are and hectare are a and ha, respectively. Thus, 5 ha is read "5 _____."

hectares

26. To change from one metric unit of area to another we again use the prefix to help us. But since we are dealing with area or square units we now square the exponential number. (Note, however, that an are, abbreviated a, and a hectare, abbreviated ha, do *not* have an exponent of 2 in their abbreviations. An are is 10^2 m^2, not $(10^2)^2$ m^2 and a hectare is 10^4 m^2, not $(10^4)^2$ m^2.)

> (a) Change the units to square meters.
> (b) Divide the given unit by the one to which you are changing.
> (c) Multiply the original number by the quotient found in Step (b).

Example: Change 3.5 mm^2 to cm^2.

(a) $mm^2 = (10^{-3})^2 \, m^2 = 10^{-6} \, m^2$
   $cm^2 = (10^{-2})^2 \, m^2 = 10^{-4} \, m^2$

(b) $\dfrac{mm^2}{cm^2} = \dfrac{10^{-6} \, m^2}{10^{-4} \, m^2} = 10^{-6 - ^-4} = 10^{-2}$

(c) $3.5 \, mm^2 = 3.5 \cdot 10^{-2} \, cm^2$
   $= .035 \, cm^2$

Change 25 cm^2 to mm^2.

(a) $cm^2 = (10^{-2})^2 \, m^2 = 10^{-4} \, m^2$
   $mm^2 = (10^{-3})^2 \, m^2 = 10^{-6} \, m^2$

(b) $\dfrac{cm^2}{mm^2} = \dfrac{10^{-4} \, m^2}{10^{-6} \, m^2} = 10^{-4 - ^-6}$

(c) $25 \, cm^2 = 25 \cdot 10^2 \, mm^2$
   $= 2,500 \, mm^2$

27. $.003 \, dkm^2 =$ _____ dm^2

(a) $dkm^2 = (10^1)^2 \, m^2 = 10^2 \, m^2$
   $dm^2 = (10^{-1})^2 \, m^2$
   $= 10^{-2} \, m^2$

(b) $\dfrac{10^2 \, m^2}{10^{-2} \, m^2} = 10^4$

(c) $.003 \, dkm^2 = .003 \cdot 10^4 \, dm^2$
   $= 30 \, dm^2$

28. This procedure does not apply in its entirety to ares and hectares since they are defined in terms of square meters rather than meters. To make these conversions you need only remember that an are is 100 m² or 10^2 m², while a hectare is 10,000 m² or 10^4 m².

Example: Convert .5 ha to square dekameters.

(a) $\text{ha} = 10^4 \text{ m}^2$
$\text{dkm}^2 = (10^1)^2 \text{ m}^2 = 10^2 \text{ m}^2$

(b) $\dfrac{\text{ha}}{\text{dkm}^2} = \dfrac{10^4 \text{ m}^2}{10^2 \text{ m}^2} = 10^2$

(c) $.5 \text{ ha} = .5 \cdot 10^2 \text{ dkm}^2 = 50 \text{ dkm}^2$

$35 \text{ a} = \underline{\hspace{4cm}} \text{ dm}^2$

(a) $\text{a} = 10^2 \text{ m}^2$
$\text{dm}^2 = (10^{-1})^2 \text{ m}^2 = 10^{-2} \text{ m}^2$

(b) $\dfrac{\text{a}}{\text{dm}^2} = \dfrac{10^2 \text{ m}^2}{10^{-2} \text{ m}^2} = 10^4$

(c) $35 \text{ a} = 35 \cdot 10^4 \text{ dm}^2$
$= 350,000 \text{ dm}^2$

29. $45 \text{ dm}^2 = \underline{\hspace{4cm}} \text{ m}^2$

(a) $\text{dm}^2 = (10^{-1})^2 \text{ m}^2 = 10^{-2} \text{ m}^2$
$\text{m}^2 = 10^0 \text{ m}^2$

(b) $\dfrac{10^{-2} \text{ m}^2}{10^0 \text{ m}^2} = 10^{-2}$

(c) $45 \text{ dm}^2 = 45 \cdot 10^{-2} \text{ m}^2$
$= .45 \text{ m}^2$

Exercise Set 9-2. Work all the problems before checking your answers.

1. $1 \text{ are} = \underline{\hspace{1.5cm}} \text{ m}^2$

2. $2 \text{ hetares} = \underline{\hspace{1.5cm}} \text{ m}^2$

3. $5 \text{ cm}^2 = \underline{\hspace{1.5cm}} \text{ m}^2$

4. $950 \text{ dm}^2 = \underline{\hspace{1.5cm}} \text{ m}^2$

5. $.8 \text{ m}^2 = \underline{\hspace{1.5cm}} \text{ cm}^2$

6. $.005 \text{ m}^2 = \underline{\hspace{1.5cm}} \text{ mm}^2$

7. $48 \text{ mm}^2 = \underline{\hspace{1.5cm}} \text{ cm}^2$

8. $85 \text{ cm}^2 = \underline{\hspace{1.5cm}} \text{ mm}^2$

9. $1,000 \text{ dm}^2 = \underline{\hspace{1.5cm}} \text{ a}$

9-3 VOLUME METRIC UNITS

In this section we shall consider metric units of volume. Although we shall not be considering many metric units, such as those used for weights, power, and heat, the techniques developed in this chapter can be used in working with them also.

30. Although the cubic meter, the volume of a cube with sides of 1 meter, is the basic unit of volume measure, it is not the most important. The cubic decimeter, called the *liter*, is used more often in liquid measure. By using the usual prefixes we can indicate parts of a liter. For example, a centiliter is 10^{-2} liter.
    1 milliliter is _____ liter 10^{-3}

31. 1 kiloliter = _____ liters 10^3

32. 1 dekaliter = _____ liters 10^1

33. Since 1 liter is equal to the volume of a cube with sides 10^{-1} meter, a liter is $(10^{-1})^3$ cubic meter or _____ cubic meter. 10^{-3}

34. The volumes cubic millimeter, cubic centimeter, cubic meter are abbreviated using an exponent of 3. For example, cubic millimeter is abbreviated mm^3. Cubic centimeter is abbreviated _____, although cc is often used instead. cm^3

35. Cubic meter is abbreviated _____. m^3

36. The abbreviation for liter is ℓ. The abbreviations for the prefixes of the liter are as in linear measure, so deciliter is abbreviated $d\ell$. Give the abbreviation for milliliter. $m\ell$

37. Give the abbreviations for the following:
    centiliter _____ $c\ell$
    dekaliter _____ $dk\ell$
    hectoliter _____ $h\ell$
    kiloliter _____ $k\ell$

38. Since a centimeter is 10^{-2} meter, a cubic centimeter is _____ cubic meter. (*Hint*: a square centimeter is $(10^{-2})^2 = 10^{-4}$ square meters.) $(10^{-2})^3$ or 10^{-6}

39. A cubic millimeter (mm^3) is _____ m^3. $(10^{-3})^3 = 10^{-9}$

40. km^3 = _____ m^3 $(10^3)^3 = 10^9$

41. Since the liter is often used as the basic unit for liquid measure, you will need to convert from centiliters to milliliters, from kiloliters to dekaliters, etc. Since the conversions are linear (there is no exponent in the abbreviations), you can use the technique outlined in Section 9-1 to convert from one to the other.

Example: Change 1.7 mℓ to centiliters.

    (a) mℓ = $10^{-3}\,\ell$; cℓ = $10^{-2}\,\ell$

    (b) $\dfrac{m\ell}{c\ell} = \dfrac{10^{-3}\,\ell}{10^{-2}\,\ell} = 10^{-3--2} = 10^{-1}$

    (c) 1.7 mℓ = $1.7 \cdot 10^{-1}$ cℓ = .17 cℓ

.41 dℓ = _____ mℓ

    (a) dℓ = $10^{-1}\ell$

        mℓ = $10^{-3}\,\ell$

    (b) $\dfrac{d\ell}{m\ell} = \dfrac{10^{-1}\ell}{10^{-3}\,\ell} = 10^2$

    (c) .41 dℓ = $.41 \cdot 10^2$ mℓ

            = 41 mℓ

42. 45 mℓ = _____ cℓ

    4.5 cℓ

43. 15 kℓ = _____ dkℓ

    1,500 dkℓ

44. If you wish to convert from one cubic unit to another, you can proceed in a manner similar to the one shown in Section 9-2. The steps are

> (a) Change the metric units to cubic meters.
> (b) Divide the given unit by the one to which you are changing.
> (c) Multiply the original number by the quotient found in Step (b).

Example: Change 30 mm^3 to cubic centimeters.

    (a) mm^3 = $(10^{-3})^3$ m^3 = 10^{-9} m^3

        cm^3 = $(10^{-2})^3$ m^3 = 10^{-6} m^3

    (b) $\dfrac{mm^3}{cm^3} = \dfrac{10^{-9}\ m^3}{10^{-6}\ m^3} = 10^{-3}$

    (c) 30 mm^3 = $30 \cdot 10^{-3}$ cm^3 = .03 cm^3

Change 145 mm^3 to cubic centimeters.

    (a) mm^3 = 10^{-9} m^3

        cm^3 = 10^{-6} m^3

    (b) $\dfrac{mm^3}{cm^3} = \dfrac{10^{-9}\ m^3}{10^{-6}\ m^3} = 10^{-3}$

    (c) 145 mm^3 = $145 \cdot 10^{-3}$ cm^3

            = .145 cm^3

45. 170 cm^3 = _____ dkm^3

    (a) cm^3 = $(10^{-2})^3$ m^3

            = 10^{-6} m^3

        dkm^3 = $(10^1)^3$ m^3 = 10^3 m^3

    (b) $\dfrac{cm^3}{dkm^3} = \dfrac{10^{-6}\ m^3}{10^3\ m^3} = 10^{-9}$

    (c) 170 cm^3 = $170 \cdot 10^{-9}$ dkm^3

            = .00000017 dkm^3

46. $.1 \text{ km}^3 = $ _____ hm^3

(a) $\text{km}^3 = (10^3)^3 \text{ m}^3 = 10^9 \text{ m}^3$
$\text{hm}^3 = (10^2)^3 \text{ m}^3 = 10^6 \text{ m}^3$

(b) $\dfrac{\text{km}^3}{\text{hm}^3} = \dfrac{10^9 \text{ m}^3}{10^6 \text{ m}^3} = 10^3$

(c) $.1 \text{ km}^3 = .1 \cdot 10^3 \text{ hm}^3$
$= 100 \text{ hm}^3$

Exercise Set 9-3.   Work all the problems before checking your answers.

1. $35 \text{ m}\ell = $ _____ ℓ

2. $900 \; \ell = $ _____ $\text{k}\ell$

3. $40 \text{ c}\ell = $ _____ $\text{m}\ell$

4. $75 \text{ k}\ell = $ _____ $\text{c}\ell$

5. $350 \text{ cm}^3 = $ _____ mm^3

6. $.49 \text{ mm}^3 = $ _____ cm^3

_____

_____

_____

_____

_____

_____

9-4   ENGLISH-METRIC CONVERSIONS

Since most of us have been using the English system all our lives, we think in terms of English units.   In order to function using the metric system we must be able to convert back and forth between the metric system and the English system.   In this section we shall look at these conversions.

---

47. I stated previously that a meter was approximately equal to the English yard.   In the United States the meter is defined to be 39.37 inches as compared with 36 inches for the yard.   In other words, 1 meter is 1.094 yards or, looking at the problem the other way, 1 yard is .9144 meter.   These are actually the only conversions you would need.   For convenience, however, a table with a number of conversion factors is provided in the back of the book.   To change from meters to feet we would multiply the number of meters by the number of feet per 1 meter.   Since 1 meter = 3.2808 feet, to change 2 meters to feet we would use the following step:

$$\frac{\overset{1}{(2 \text{ meters})}}{1} \frac{(3.2808 \text{ feet})}{\underset{1}{\text{meter}}} = 6.5616 \text{ feet}$$

3 meters = _____ feet

$$\frac{(3 \text{ m})(3.2808 \text{ ft/m})}{1} = \underline{9.8424 \text{ ft}}$$

48. There are 1.094 yards per meter so
7 meters = _____ yards.

$$\frac{(7 \text{ m})(1.094 \text{ yd/m})}{1} = \underline{7.658 \text{ yd}}$$

49. There are .6214 mile per kilometer so

    11 kilometers = _____ miles.

$$\frac{1}{(11 \text{ km})(.6214 \text{ mi/km})}$$
$$= \underline{6.8354 \text{ miles}}$$

50. There are 39.37 inches per meter so

    5 meters = _____ inches.

$$\frac{1}{(5 \text{ m})(39.37 \text{ in./m})} = \underline{196.85 \text{ inches}}$$

51. To change 55 cm to inches we need only to convert 55 cm to meters as shown in Section 9-1 and then proceed as above.

    55 cm = _____ m

    .55

52. 55 cm = .55 m = _____ inches

$$\frac{1}{(.55 \text{ m})(39.37 \text{ in./m})}$$
$$= \underline{21.6535 \text{ inches}}$$

53. To change a metric unit to an English unit the following steps are used.

(a) Find the English unit in the "Metric to English" portion of the table and, if the metric unit is *not* the one given in the table, change the metric unit to the one given in the table.
(b) Multiply the number of metric units found in Step (a) by the conversion factor given in the table.

    Example: Change 35 hm to miles.

    (a) Since the table gives the conversion factor for kilometers, not hectometers, change 35 hm to kilometers.

    35 hm = 3.5 km

    (b) $\dfrac{1}{(3.5 \text{ km})(.6214 \text{ mi/km})} = 2.1749 \text{ mi}$

    Change .8 dm to yards.

    (a) .8 dm = .08 m

    (b) $\dfrac{1}{(.08 \text{ m})(1.094 \text{ yd/m})}$
    $= .08752 \text{ yd}$

54. .11 dm² = _____ in.²

    (a) .11 dm² = 11 cm²

    (b) $\dfrac{1}{(11 \text{ cm}^2)(.155 \text{ in.}^2/\text{cm}^2)}$
    $= 1.705 \text{ in.}^2$

55. 13 a = _____ acre

$$\frac{1}{(13\ \cancel{a})(.0247\ acre/\cancel{a})} = .3211\ acre$$

56. .4 m² = _____ ft²

$$\frac{1}{(.4\ \cancel{m^2})(10.764\ ft^2/\cancel{m^2})} = 4.3056\ ft^2$$

57. 7 mm² = _____ in.²

(a) 7 mm² = .07 cm²

(b) $\dfrac{1}{(.07\ \cancel{cm^2})(.155\ in.^2/\cancel{cm^2})}$

    = .01085 in.²

58. .9 m³ = _____ yd³

(b) (.9 m³)(1.308 yd³/m³)
     = 1.177 yd³

59. 30 ℓ = _____ lb of water at 62°F

(b) (30 ℓ)(2.202 lb/ℓ) = 66.06 lb

60. 350 mm³ = _____ in.³

(a) 350 mm³ = .35 cm³
(b) (.35 cm³)(.061 in.³/cm³)
     = .021 in.³

61. To convert from the English system to the metric system, use the "English to Metric" portion of the table and multiply the number of English units by the number of metric units per 1 English unit. Thus,

$$5\ yards = 4.572\ meters\ since\ \frac{1}{(5\ \cancel{yd})(.9144\ m/\cancel{yd})} = 4.572\ meters.$$

12 yards = _____ meters

$$\frac{1}{(12\ \cancel{yd})(.9144\ m/\cancel{yd})}$$
$$= \underline{10.97\ meters}$$

(*Note*: From this point on the answers are rounded off to 4 or 5 significant digits in accordance with the rules of Chapter 2 on significant digits since the conversion factors are only accurate to either 4 or 5 places.)

62. To convert from English units to metric units the following steps are used:

(a) Change from the English unit to the metric unit given in the table by multiplying the number of English units by the conversion factor given in the table.

(b) Change the answer found in Step (a) to the desired metric unit by the procedure outlined in Section 9-1.

Example: Change 13 feet to centimeters.

(a) $(13 \text{ ft})(.3048 \text{ m/ft}) = 3.962 \text{ m}$

(b)   Change 3.962 m to centimeters.
     $3.962 \text{ m} = 396.2 \text{ cm}$

2 feet = _____ centimeters

(a) $(2 \text{ ft})(.3048 \text{ m/ft}) = .6096 \text{ m}$

(b) $.6096 \text{ m} = 60.96 \text{ cm}$

63.   25 inches = _____ meter

(a) $(25 \text{ in.})(25.4 \text{ mm/in.}) = 635 \text{ mm}$

(b) $635 \text{ mm} = .635 \text{ m}$

64.   .4 miles = _____ meters

$(.4 \text{ mi})(1.069 \text{ km/mi})$
$= .6436 \text{ km} = \underline{643.6 \text{ m}}$

65.   3 yards = _____ dekameter

$(3 \text{ yd})(.9144 \text{ m/yd})$
$= 2.743 \text{ m} = \underline{.2743 \text{ dkm}}$

66.   9 acres = _____ a

$(9 \text{ acres})(40.47 \text{ a/acre}) = 364.2 \text{ a}$

67. $15 \text{ yd}^2 = $ _____ m^2

$(15 \text{ yd}^2)(.836 \text{ m}^2/\text{yd}^2) = 12.5 \text{ m}^2$

68. $30 \text{ in.}^2 = $ _____ cm^2

$(30 \text{ in.}^2)(6.452 \text{ cm}^2/\text{in.}^2)$
$= 193.6 \text{ cm}^2$

69. $2 \text{ ft}^2 = $ _____ cm^2

(a) $(2 \text{ ft}^2)(.0929 \text{ m}^2/\text{ft}^2) = .186 \text{ m}^2$

(b) $.186 \text{ m}^2 = 1,860 \text{ cm}^2$

70. $200 \text{ in.}^3 = $ _____ cm^3

(a) $(200 \text{ in.}^3)(16.383 \text{ cm}^3/\text{in.}^3)$
$= 3,276.6 \text{ cm}^3$

71. 200 in.^{3} = _____ ℓ

(a) 200 in.^{3} = 3,276.6 cm^3
(See frame 70)

(b) 3,276.6 cm^3 = 3.2766 ℓ
  (i) cm^3 = 10^{-6} m^3
    ℓ = dm^3 = 10^{-3} m^3

  (ii) $\dfrac{10^{-6}\,\text{m}^3}{10^{-3}\,\text{m}^3} = 10^{-3}$

  (iii) 3,276.6 cm^3
    = 3,276.6 · 10^{-3} ℓ
    = 3.2766 ℓ

72. 30 in.^{3} = _____ dℓ

(a) (30 in.^{3})(16.383 cm^3/in.^{3})
  = 491.49 cm^3

(b) (i) cm^3 = $(10^{-2})^3$ m^3
    = 10^{-6} m^3

    dℓ = 10^{-1} ℓ
    = 10^{-1} dm^3
    = $10^{-1} \cdot (10^{-1})^3$ m^3
    = $10^{-1} \cdot 10^{-3}$ m^3
    = 10^{-4} m^3

  (ii) $\dfrac{\text{cm}^3}{\text{d}\ell} = \dfrac{10^{-6}\,\text{m}^3}{10^{-4}\,\text{m}^3} = 10^{-2}$

  (iii) 491.49 cm^3
    = 491.49 · 10^{-2} dℓ
    = 4.9149 dℓ

Exercise Set 9-4. Work all the problems before checking your answers.

1. 17 meters = _____ feet
2. 25 kilometers = _____ mile
3. 50 dekameters = _____ yards
4. 30 millimeters = _____ inch
5. 22 inches = _____ millimeters
6. 14.3 miles = _____ kilometers
7. 1,000 yards = _____ kilometer
8. 1.9 feet = _____ centimeters
9. 150 cm^2 = _____ in.^{2}
10. 50 a = _____ acres
11. 3 ha = _____ acres
12. 49 mm^2 = _____ in.^{2}
13. .04 m^2 = _____ in.^{2}
14. 1 acre = _____ ha
15. 35 yd^2 = _____ m^2
16. 47 in.^{2} = _____ mm^2

17. $3 \text{ yd}^2 =$ _____ dm^2 _____

18. $50 \text{ ft}^2 =$ _____ m^2 _____

19. $2000 \text{ cm}^3 =$ _____ in.^3 _____

20. $5 \ell =$ _____ gal   _____

21. $50 \text{ mm}^3 =$ _____ in.^3 _____

22. $30 \ell =$ _____ yd^3 _____

23. $.05 \text{ m}^3 =$ _____ in.^3 _____

24. $9 \text{ m}\ell =$ _____ qt   _____

25. $15 \text{ qt} =$ _____ ℓ _____

26. $90 \text{ ft}^3 =$ _____ m^3 _____

27. $300 \text{ in.}^3 =$ _____ mm^3 _____

28. $490 \text{ qt} =$ _____ $\text{k}\ell$ _____

9-5 MAKING FRIENDS WITH THE METRIC SYSTEM

When learning a foreign language an individual has not really mastered the language until he starts thinking in the new language, not thinking in the native language and then translating into the new language. The same is true about mastering the metric system. After learning to translate (convert) metric to English and English to metric you must then learn to think in metric terms. The purpose of this section is to help you take the first step toward thinking metric by helping you estimate the lengths, areas, and volumes of familiar objects in the metric units.

73. The basic unit of the metric system is the meter, which is approximately equal to a yard. The normal stride is approximately one meter. How high would you guess an average door is? (Express your answer in meters.)   2 meters

74. If the average ceiling is 8 feet high, guess the approximate height in meters. $2\frac{1}{2}$ meters

75. A decimeter is one-tenth of a meter. The average hand is approximately a decimeter wide. Using this as a guide, what are the approximate dimensions of a brick in decimeters? $\frac{1}{2} \text{ dm} \times 1 \text{ dm} \times 2 \text{ dm}$

76. A millimeter is often used for wrench sizes. For example, $\frac{3}{4}''$ is almost exactly 19 mm. Try to guess the closest metric wrench to $\frac{1}{4}''$.   6 mm

77. Guess the closest metric wrench to $\frac{1}{2}''$.   13 mm

78. A kilometer is used to measure great distances (that in the English system are usually measured in miles) in the metric system. A kilometer is approximately $\frac{5}{8}$ of a mile. Thus, two cities 50 miles apart are approximately _____ kilometers apart.   80

79. If the speed limit is 55 miles per hour, how many kilometers per hour is that?

88

80. The length of the piston stroke of a car is usually between 6 and 10 centimeters. Guess what the bore (diameter of the cylinder) will usually be.

between 6 and 10 centimeters

81. An event in track and field is the 1,500-meter run. Why is this distance used? 1,500 meters is approximately equal to what English unit?

the mile

82. The 400-meter run is approximately equal to what distance in miles?

$\frac{1}{4}$ mile

83. If we look at area measures, a door is approximately 2 square meters. What is the approximate area of a sheet of paneling (4′ × 8′)?

3 m^2

84. What is the approximate area of a 12′ × 16′ room? (Compare it with a 4′ × 8′ sheet of paneling.)

18 m^2 (since it is 6 times as large as a 4′ × 8′ sheet).

85. Figure 9-1 shows one square centimeter. Using it as a guide, what would you guess is the area of the nail head to the right of the figure?

$\frac{1}{3}$ cm^2

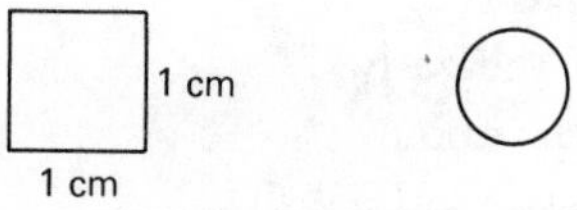

FIGURE 9-1

86. What would you guess is the area of the piece shown in Fig. 9-2?

6 cm^2

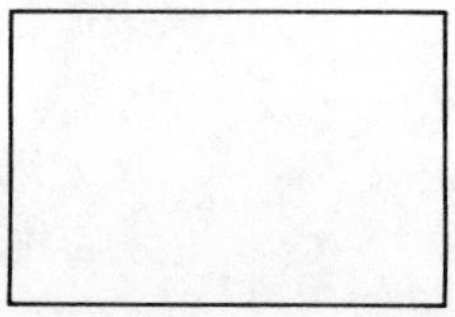

FIGURE 9-2

87. A square millimeter is very small. The area of the head of a straight pin is approximately 1 mm^2. What is the area of a period (.) in square millimeters?

approximately $\frac{1}{4}$ mm^2

88. In volume measure the liter and cubic centimeter are used quite often. The liter and the quart are approximately equal. Consider a large gas tank (24 gallons). Approximately how many liters does it contain?

100 ℓ or 1 hℓ

89. A 4-cylinder engine may have a piston displacement of from 1 to 3 liters. What displacements do these engines have in cubic centimeters?

1,000 to 3,000 cc

90. A cubic yard is approximately $\frac{3}{4}$ of a cubic meter. If you need 8 cubic yards of concrete, how many cubic meters should you order?

6 m³

91. A cubic millimeter is about the size of a broken off pencil point. Approximately what is the volume of a drop of solder?

20 to 30 mm³

Exercise Set 9-5.   Work all the problems before checking your answers.

1. What is the approximate length and width of the rectangle in Fig. 9-3?

_____

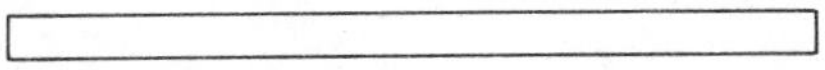

FIGURE 9-3

2. What is the approximate area of the rectangle in Fig. 9-3 in
   (a) square millimeters?
   (b) square centimeters?

_____

_____

3. Name something that has a volume of approximately 2 liters.

_____

SELF-TEST

1. Define each prefix below:
   (a) milli
   (b) kilo
   (c) hecto

   _____

   _____

   _____

2. 30 dkm = _____ dm

   _____

3. 100 mm = _____ km

   _____

4. 35 km = _____ mi

   _____

5. .87 foot = _____ cm

   _____

6. 5 hectare = _____ m²

   _____

7. 35 cm² = _____ m²

   _____

8. 150 cm² = _____ mm²

   _____

9. 2,000 dm² = _____ ha

   _____

10. 25 cm² = _____ in.²

    _____

11. .04 m² = _____ in.²

    _____

12. 1 acre = _____ ha

    _____

13. 15 yd² = _____ dm²

    _____

14. 905 dℓ = _____ ℓ

    _____

15. 29 cm³ = _____ mm³

    _____

16. 490 mm³ = _____ cm³

    _____

17. 3200 cm^3 = _____ in.^{3}

18. 30 ℓ = _____ qt

19. .8 qt = _____ dℓ

20. 90 ft^3 = _____ m^3

21. 7 mℓ = _____ qt

22. 900 mm^3 = _____ in.^{3}

23. What is the approximate volume of a small (12 gallon) gas tank?

24. Estimate the volume of a deck of cards if 6 cubic inches is
     approximately 100 cubic centimeters.

25. Estimate the area of an 8$\frac{1}{2}$″ × 11″ sheet of paper in square
     centimeters and in square decimeters.

_____ ; _____

APPLIED PROBLEMS

In each of the following problems one of the variables X, Y, Z, etc., associated
with the original problem is missing from the table. Use the concepts of this
chapter to find the missing variable.

A. X kilometers per hour and Y miles per hour are equivalent.

	X	Y
1.	120	
2.	80.6	
3.		55
4.		48.3
5.	91.3	

Example: 1. We must change 120 kilometers to miles. The conversion
factor is .621 mi/km, so 120 km = (120 km)(.621 mi/km) = 74.5 miles.
Thus, 120 km/hr ≈ 74.5 mph.

B. Bore dimensions of X inches and Y millimeters are equivalent.

	X	Y
1.	4	
2.	3.83	
3.		98
4.		94.7

C. Turn diameters of X feet and Y meters are equivalent.

	X	Y
1.	14.7	
2.	$20\frac{1}{3}$	
3.		4
4.		4.1

D. A total piston displacement of X liters, Y cubic centimeters, and Z cubic inches are all equivalent.

	X	Y	Z
1.	3.2		
2.		2,300	
3.	4.3		
4.			145
5.		3,600	
6.			230
7.	4.8		

Example: 1. 3.2 liters $= 3.2 \text{ dm}^3 = 3,200 \text{ cm}^3$

$3,200 \text{ cm}^3 = (3,200 \text{ cm}^3)(.061 \text{ in.}^3/\text{cm}^3) \approx 195 \text{ in.}^3$

E. A metric wrench of X millimeters is closest to an American wrench of Y inches.

	X	Y
1.	14	
2.	12	
3.		$\frac{1}{2}$
4.		$\frac{3}{4}$
5.	11	
6.	16	
7.		$\frac{7}{8}$
8.	10	

Example: 1. 14 mm $\approx .5516''$. The closest American wrench size is $\frac{9}{16}'' = .5625''$.

F. *X* gallons per minute and *Y* liters per minute are equivalent.

	X	*Y*
1.	400	
2.		850
3.	15.2	
4.		1200
5.	257	

G. An air sample contains *W* micrograms of particle per cubic meter. This is equivalent to *X* micrograms per liter, *Y* micrograms per cubic dekameter, and *Z* micrograms per cubic centimeter.

	W	*X*	*Y*	*Z*
1.	725			
2.		.61		
3.			598,000	
4.				.0011
5.	835			

H. The shop building was *X* feet wide by *Y* feet long.

	X	*Y*	ft^2	m^2	a
1.	45	400			
2.	90	120			
3.	100	221			
4.	42	55			
5.	108	320			

I. In the text material the metric units of mass were not discussed. The basic unit of mass in the metric system is the gram, just as the meter is the basic unit of length. The gram was originally defined as the mass of a cubic centimeter of pure water at 4°C but has since been redefined in terms of a standard, a platinum cylinder defined to have a mass of 1,000 grams—a kilogram. Since the mass of an object and its weight are almost identical everywhere on the earth, the gram is generally used as a measure of weight.

The abbreviation for the gram is g. Using the prefixes and techniques developed in the chapter, convert the following mass units to the specified unit. (*Note*: A metric ton is 1,000 kilograms.)

1. 3 mg = _____ g
2. 7.5 kg = _____ hg
3. .04 cg = _____ mg
4. .29 mg = _____ g
5. 428 dkg = _____ kg
6. 25 kg = _____ metric ton
7. 3.2 metric tons = _____ kg
8. 638 mg = _____ kg

J. Below is a table of conversion factors to be used when dealing with weights. Use it to make the conversions indicated below. (*Note*: A ton is 2,000 pounds.)

Metric to *English*: *Conversion Factor*
Grams to Pounds: .0022 lb/g
Grams to Ounces: .0353 oz/g
Metric Tons to Pounds: 2,200 lb/m tons
Metric Tons to Tons: 1.1 tons/m tons
Kilograms to Pounds: 2.2 lb/kg

English to *Metric*: *Conversion Factor*
Pounds to Grams: 453.6 g/lb
Ounces to Grams: 28.35 g/oz
Pounds to Metric Tons: .000454 m ton/lb
Tons to Metric Tons: .907 m ton/ton
Pounds to Kilograms: .4536 kg/lb

1.   428 g = _____ lb
2.   38 kg = _____ lb
3.   4 lb = _____ kg
4.   3 oz = _____ g
5.   .04 oz = _____ mg
6.   .2 lb = _____ mg
7.   73 mg = _____ oz
8.   3 tons = _____ metric tons
9.   32.6 metric tons = _____ tons
10. 4 lb 3 oz = _____ g
11. 1 lb 13 oz = _____ mg
12. 40 lb 2 oz = _____ hg

K. Using the conversion factors in Problem J as a guide, estimate the metric weight of the following objects:

	Object	Metric Weight
1.	a mid-size automobile	
2.	a 10-ounce glass of water	
3.	a 150-pound man	
4.	a nail	
5.	a brick	
6.	a gallon of milk	

10

ratio and proportion (part 2)

OBJECTIVES

After completing this chapter the student should be able to
1.   Solve word problems that require the use of an inverse proportion.
2.   Solve word problems that require the use of a compound proportion.

(1–7)   **1.**   If two gears are in mesh and gear 1 has 45 teeth and gear 2 has 25 teeth with gear 1 turning 300 rpms, how fast is gear 2 turning?

                                                            _____

(8–13)   **2.**   If 6 men can resurface 200 yards of road in 4 hours, how many yards can 10 men resurface in 9 hours?

                                                            _____

10-1   INVERSE PROPORTIONS

Recall from Chapter 6 that the gear with the fewer number of teeth turns faster than the gear with more teeth. If gear A has 20 teeth and gear B has 10 teeth and gear A is turning at 200 rpms, then gear B must be turning at twice that rate (since the speed ratio is 2:1) or 400 rpms. If we try to set up a direct proportion as

$$\frac{\text{No. of teeth in gear A}}{\text{No. of teeth in gear B}} = \frac{\text{rpms of gear A}}{\text{rpms of gear B}} \quad \text{or} \quad \frac{20 \text{ teeth}}{10 \text{ teeth}} = \frac{200 \text{ rpms}}{N}$$

where N is the number of rpms of gear B, however, we find an incorrect answer of 100 rpms. This is because the relationship between the number of teeth and the rate is an inverse relationship; that is, as the number of teeth on one gear becomes larger while the number of teeth on the other gear remains unchanged, the rate of the first gear becomes smaller. To illustrate this, look at the following speed ratios.

gear A = 20 teeth   speed ratio = 2:1
gear B = 10 teeth   (B to A)

gear A = 20 teeth   speed ratio = 1.25:1
gear B = 16 teeth   (B to A)

Notice that the speed ratio becomes smaller when the number of teeth in gear B increases with gear A unchanged. We say that the ratio of the number of teeth and the speed ratio are *inversely proportional*.

---

1. An **inverse proportion** is one in which one of the ratios of like units is equal to the reciprocal of the other ratio that would be used in a direct proportion. For example, in the problem above the (incorrect) direct proportion

$$\frac{20 \text{ teeth}}{10 \text{ teeth}} = \frac{200 \text{ rpms}}{N}$$

can be changed to an inverse proportion by inverting one of the ratios. Change the proportion to an inverse proportion.

           $\dfrac{20}{10} = \dfrac{N}{200}$ or $\dfrac{10}{20} = \dfrac{200}{N}$

2. An inverse proportion is solved in exactly the same way as a direct proportion. Solve

$$\frac{20}{10} = \frac{N}{200}$$

           $(200)\dfrac{20}{10} = (200)\dfrac{N}{200}$

           $400 = N$

3. Since an inverse proportion states that the ratios are equal, the units must be the same. Since the one ratio will be the reciprocal of the corresponding direct proportion ratio, this can only be accomplished if each ratio in the proportion has the same units for both the numerator and denominator. For this reason when using an inverse proportion the quantities with the same units must be put in the same ratio. Is

$$\frac{20 \text{ teeth}}{400 \text{ rpms}} = \frac{500 \text{ rpms}}{N \text{ teeth}}$$

an inverse proportion?

No, since the units (teeth/rpms and rpms/teeth) are different. The proportion would have to be written

$$\frac{20 \text{ teeth}}{N \text{ teeth}} = \frac{500 \text{ rpms}}{400 \text{ rpms}}$$

4. If gear A has 18 teeth and gear B has 10 teeth with gear A turning 300 rpms, what is the rpm of gear B? (Use an inverse proportion to solve the problem.)

540 rpms

$$\frac{18 \text{ teeth (gear A)}}{10 \text{ teeth (gear B)}}$$

$$= \frac{N \text{ (gear B)}}{300 \text{ rpms (gear A)}}$$

$$(300)\frac{18}{10} = (300)\frac{N}{300}$$

$$540 = N$$

5. If you are unsure about whether to use a direct or inverse proportion to solve a particular problem, it is usually best to set up a direct proportion to see if the resulting answer is reasonable. For example, consider a system of pulleys. If the distance around the outside of one pulley (called the *circumference*) is 25 inches and the circumference of the second pulley is 15 inches, with the larger pulley turning at 300 rpms, what is the rate of the smaller pulley? A direct proportion to solve this problem would be

$$\frac{25''}{15''} = \frac{300 \text{ rpms}}{N \text{ rpms}}$$

where the 300 rpms is put on the top of the second ratio because it is the rate of the 25″ pulley that is in the numerator of the first ratio. Solve this proportion for N.

$$(15N)\frac{25}{15} = (15N)\frac{300}{N}$$

$$25 \times N = 4,500$$

$$N = 180 \text{ rpms}$$

6. Refer to frame 5. When the larger pulley has made one complete revolution, the pulley belt has moved 25 inches. Thus, the smaller pulley will have moved through 25 inches also, so it will have made $1\frac{2}{3}$ revolutions. Thus, the smaller pulley must be turning at a rate faster than the larger pulley. Therefore, the direct proportion used in frame 5 must be incorrect. Solve the problem using an inverse proportion.

$$\frac{25''}{15''} = \frac{N}{300 \text{ rpms}}$$

$$(300)\frac{25}{15} = (300)\frac{N}{300}$$

$$500 = N$$

7. If pulley 1 has a diameter of 12″ and pulley 2 has a diameter of 8″ and pulley 1 is turning 350 rpms, what is the rate of pulley 2?

525 rpms

$$\frac{12'' \text{ (pulley 1)}}{8'' \text{ (pulley 2)}} = \frac{N}{350 \text{ rpm (pulley 1)}}$$

$$(1{,}400)\frac{12}{8} = (1{,}400)\frac{N}{350}$$

$$2{,}100 = 4N$$

$$525 = N$$

Exercise 10-1. Work all the problems before checking your answers.

1. If gear A has 30 teeth and gear B has 20 teeth with gear A turning 350 rpms, how fast is gear B turning?

2. If two pulleys are on the same belt and the circumference of pulley 1 is 32 inches and the circumference of pulley 2 is 18 inches with pulley 1 turning 500 rpms, how fast is pulley 2 turning?

10-2 COMPOUND PROPORTIONS

If more than two ratios are involved in solving a problem by using a proportion, a compound proportion must be used. For example, the problem "If 8 men can machine 35 pieces in 2 days, how many pieces can 7 men machine in 5 days?" can be solved using a compound proportion.

---

8. Consider the problem stated above. There are two cases in the problem. In the first case, case A, 8 men machine 35 pieces in 2 days.

In the second case, case B, 7 men machine N pieces in 5 days. The obvious ratios are 8 men/7 men, 35 pieces/N pieces, and 2 days/5 days. Since the number of pieces in case B is unknown, the ratio on one side of the proportion will be 35/N, where the 35 came from case A and the N from case B. Next, consider the ratio 8 men/7 men. Is it directly or inversely related with the number of pieces machined? That is, if you increase the number of men, will more pieces be machined? If so, the relation is direct. If not, it is an inverse relation.

Directly related, since increasing the number of men *increases* the number of pieces machined.

9. Refer to frame 8. Since the number of men is directly proportional to the number of pieces machined, the ratio should be in the same order as the ratio of pieces machined. Since 35/N is case A over case B, the ratio of men should be case A over case B. Since there are 8 men in case A and 7 men in case B, the ratio is _____.

8 men/7 men

10. Refer to frames 8 and 9. Consider the number of days. If you increase the number of days, does the number of pieces machined increase?

yes

11. Thus, the number of days is directly proportional to the number of pieces machined, so the ratio should be in the same order as 35/N. Write the ratio of days in the correct order.

2 days/5 days
(They worked 2 days in case A and 5 days in case B, so the direct ratio of case A to case B is $\frac{2}{5}$.)

12. Since the number of men as well as the number of days worked affects the number of pieces machined, the total effect can be found by multiplying the ratios. Thus,

$$\frac{8}{7} \times \frac{2}{5} = \frac{35}{N}$$

is the compound proportion to use in solving the problem. Solve this proportion. (Round off the answer to two significant digits.)

$\dfrac{16}{35} = \dfrac{35}{N}$

$16N = 35 \cdot 35$

$N = \dfrac{1,225}{16} \approx 77$ pieces

13. The following general procedure for solving problems requiring a compound proportion is derived from frames 8–12.

(a)   Determine the values in the two cases given.

(b)   Write the ratio involving the unknown directly as case A over case B and use it on one side of the proportion.

(c)   Write each ratio directly related with the unknown as case A over case B.

(d)   Write each ratio inversely related with the unknown as case B over case A.

(e)   Multiply the ratios formed in Steps (c) and (d).

(f)   Solve the resulting proportion.

Example: If 4 men can lay 2,000 bricks in 2 days, how long will it take 3 men to lay 7,500 bricks?

(a)   Case A: 4 men, 2,000 bricks, 2 days
   Case B: 3 men, 7,500 bricks, N days

(b) $\dfrac{2 \text{ days (A)}}{N \text{ days (B)}}$

(c), (d) $\dfrac{3 \text{ men (B)}}{4 \text{ men (A)}}, \dfrac{2{,}000 \text{ (A)}}{7{,}500 \text{ (B)}}$

   since increasing the number of men decreases the days, while increasing the number of bricks increases the number of days.

(e) $\dfrac{3}{4} \times \dfrac{2{,}000}{7{,}500} = \dfrac{2}{N}$

   $\dfrac{1}{5} = \dfrac{2}{N}$

(f) $1 \times N = 5 \times 2$
   $N = 10 \text{ days}$

If 3 men can paint 2 houses in 14 days, how long will it take 5 men to paint 5 houses?

(a)   Case A: 3 men, 2 houses, 14 days
   Case B: 5 men, 5 houses, N days

(b)–(e) $\dfrac{5\text{(B)}}{3\text{(A)}} \times \dfrac{2\text{(A)}}{5\text{(B)}} = \dfrac{14}{N}$

   $\dfrac{2}{3} = \dfrac{14}{N}$

(f) $2N = 42$
   $N = 21 \text{ days}$

Exercise Set 10-2.   Work all the problems before checking your answers.
Solve the following problem using a compound proportion.

1. If 2 men can install 6 cabinets in 8 hours, how long will it take 3
    men to install 15 cabinets?

_____

SELF-TEST

1. If gears A and B are in mesh, gear A has 46 teeth, gear B has 22 teeth,
    and gear A is turning 280 rpms, how fast is gear B turning?

_____

2. If pulleys 1 and 2 are on the same belt, pulley 1 has a circumference of
    40 inches, pulley 2 has a circumference of 28 inches, and pulley 1 is
    turning 330 rpms, how fast is pulley 2 turning?

_____

3. If 10 feet of a 2-inch by 4-inch metal bar weighs 80 pounds, how much
    will 8 feet of a 3-inch by 5-inch bar of the same material weigh?

_____

APPLIED PROBLEMS

A. Two pulleys, one of diameter W inches and the other of diameter X inches,
    are attached to the same belt. If the pulley of diameter W inches is turning
    Y rpms, the other pulley is turning Z rpms. Find the missing values in
    the table below. (*Hint*: Use an inverse proportion.)

	W	X	Y	Z
1.	8	4	230	
2.		8	250	400
3.	8	3		770
4.	14		200	650
5.	16	12	500	

B. If three pulleys of diameters X, Y, and Z inches are attached to the same
    belt and the rpms of the pulley of diameter X is R, then the rpms of the
    pulleys of diameter Y and Z are S and T, respectively. Find the missing
    values in the table below.

	X	Y	Z	R	S	T
1.	20	14	12	140		
2.	24	16		150		300
3.	20		14		400	500
4.	22	20	18		250	
5.	26	18	8			600

C. In Fig. 10-1, pulleys 2 and 3 are keyed on the same shaft and thus have the same rate. Find the missing values in the table if R is the rpms for pulley 1 and S is the rpms for pulley 4.

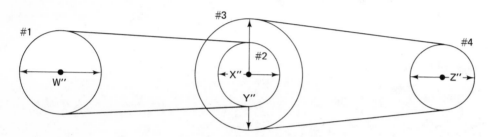

FIGURE 10-1

	W	X	Y	Z	R	S
1.	26	12	24	10	120	
2.	22	8	26	12		530
3.	16	8	20	10	135	
4.	24	9	22		100	450
5.		12	18	14	185	650
6.	28	12	18	10	140	
7.	30	15	23	12		720

Example: 1. Using a compound proportion as outlined in this chapter, we find

$$\frac{26}{12} \times \frac{24}{10} = \frac{S}{120}$$

Solving for S, we find $S = 624$ rpms.

D. If the friction loss using a W-inch diameter hose is X pounds per square inch, the friction loss is Y pounds per square inch using a Z-inch diameter hose. Find the missing values in the table below. (*Hint*: Friction loss is inversely proportional to the diameter to the fifth power of the hose.)

	W	X	Y	Z
1.	2	70		3
2.	5		80	$2\frac{1}{2}$
3.	3	24	60	
4.		12	90	2
5.	$2\frac{1}{2}$	60		3

Example: 1. $\dfrac{70}{Y} = \dfrac{(3)^5}{(2)^5}$,

so

$$\frac{70}{Y} = \frac{243}{32} \text{ and } Y \approx 9 \text{ psi.}$$

E.   Refer to Figure 10-2. Use inverse proportions to find the missing values in the table below if gear A has X teeth, gear B has Y teeth, gear C has Z teeth, gear A is turning R rpms, gear B is turning S rpms and gear C is turning T rpms.

FIGURE 10-2

	X	Y	Z	R	S	T
1.	30	15	40	1000		
2.	40	20	36		500	
3.	48	22	36			850
4.		16	30	100	200	
5.			42	200	300	150
6.	45	25			1000	400
7.		18	32	500		400
8.		20		750	1400	600

F. Refer to Figure 10-3. Use inverse proportions to find the missing values in the table below, if a force of X pounds on the brake pedal (point A) results in a force of Y pounds exerted on the master cylinder.

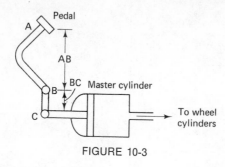

FIGURE 10-3

	AB	BC	X	Y
1.	18	2	60	
2.		4	40	300
3.	16		35	175
4.	17	4		250
5.	19	3	50	

G. Refer to Figure 10-4, a diagram of a transmission. If the crankshaft (and thus gear 1 also) is turning at R rpms, gear 1 has W teeth, gear 2 has X

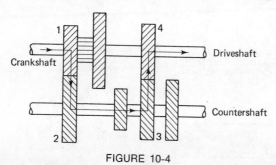

FIGURE 10-4

	R	S	W	X	Y	Z
1.	3000		20	30	14	26
2.		1200	18	36	20	30
3.	2600		24	34	16	24
4.		1000	20	28	18	30
5.	3000	1000	16	24	15	
6.	2400		22	32	15	25

teeth, gear 3 has Y teeth and gear 4 has Z teeth, then gear 4 (as well as the driveshaft) is turning S rpms. Use compound proportions to find the missing values in the table.

Example: 1. There are two reductions in rpms. One from gear 1 to gear 2 and a second from gear 3 to gear 4. Let case A contain the number of teeth in the first gear in each instance and case B the number of teeth in the second gear in each instance. Since gear 1 is turning R rpms, R rpms is in case A also, while S rpms is in case B because it is the rate of gear 4. Thus, case A is 3000 rpms, 20 teeth and 14 teeth, while case B is S rpms, 30 teeth and 26 teeth. The ratio with the unknown is 3000(A)/S(B). Since this speed ratio is inversely related to the number of teeth, the compound proportion is:

$$\frac{30(B)}{20(A)} \times \frac{26(B)}{14(A)} = \frac{3000(A)}{S(B)}$$

Solving for S gives $S \approx 1,077$.

H. In Problem G you saw how a compound proportion could be used to find the rpms of the drive shaft if you knew the rpms of the crankshaft and the number of teeth of the gears in mesh. At the rear axle the rpms are also reduced because the small drive shaft gear or pinion drives or turns the larger ring gear. Thus, to find the total gear reduction you need to know the number of teeth in the pinion and ring gear. Find the rpms that the ring gear is turning in each case below by using a compound proportion. (Refer to Fig. 10-4.)

	Crankshaft	*Gear 1*	*Gear 2*	*Gear 3*	*Gear 4*	*Pinion*	*Ring Gear*
1.	2,000	20	30	15	35	24	96
2.	1,800	24	32	18	30	26	90
3.	2,200	18	28	16	24	28	94
4.	2,400	22	34	20	32	22	88
5.	2,100	26	38	20	30	20	84
6.	1,950	24	36	22	30	22	92

11

applied geometry

OBJECTIVES

After completing this chapter the student should be able to
1. Define the terms *angle, right angle, perpendicular, acute, obtuse, parallel, rectangle, square, triangle, circle, radius, diameter, trapezoid, rectangular solid, cube, sphere.*
2. State the dimension of a geometric object.
3. Find the area of a rectangle, square, triangle, circle, trapezoid.
4. Find the volume of a cube, rectangular solid, sphere, cylinder.
5. State the Pythagorean theorem.
6. Use the Pythagorean theorem to solve various problems.

(1–31)   **1.**   (a) Two lines perpendicular to the same line at different points
                          are said to be _____.

                  (b) A point has _____ dimensions.

(32–46)   **2.**   Give a formula for the area of a triangle.

(47–51)   **3.**   Find the volume of a sphere with a radius of 10 inches.

(52–56)   **4.**   Find the hypotenuse, *c*, of a right triangle if the other sides
                  are 4 inches and 5 inches long.

11-1   DEFINITIONS

We shall be looking at two areas of applied geometry—plane geometry and solid
geometry. Plane geometry is sometimes called Euclidean geometry because the
Greek mathematician Euclid was responsible for organizing much of the theory of
plane geometry. His treatment of the subject begins with certain basic assumptions
and definitions. We shall do the same.

1. The term **dimension** refers to the concept of a geometric object taking
    up space. All real objects in our world have three dimensions—
    sometimes called *length*, *width*, and *height*. Thus, an ice cube has
    _____ dimensions.   3

2. Theoretically, it would be possible to flatten a cube until it had no
    height. Such an object would have only length and width and would
    have _____ dimensions.   2

3. If this figure was pressed together until it had no width but only
    length, we obtain a figure with _____ dimension.   1

4. Finally, if we shrink this one-dimensional figure until it has no length,
    we obtain a figure with _____ dimensions.   0

5. There are several undefined concepts that we shall assume that
    everyone understands without definition. One of these is a **point**, a
    geometric object with no length, width, or height, an object of 0
    dimensions. We represent a point by a dot, ·, but the dot is not
    actually a point since every object in the real world has _____   3
    dimensions.

6. A point is a concept used to illustrate position. The concept used to illustrate length is that of a **line**. A line has length but no width or height. Thus, it has _____ dimension.

    1

7. A line goes on forever in both directions. If we cut the line to obtain a piece of the line, we have a line **segment**. Figure 11-1 represents a line _____ since it represents a piece of a line.

    segment

FIGURE 11-1

8. A figure that has length and width but no height is called a **plane**. The surface of this paper represents a _____ figure.

    plane

9. Finally, an object with three dimensions is called a **solid**. A baseball is an example of a _____ because it has three dimensions.

    solid

10. The figure formed by two lines extending from the same point or two planes diverging from the same line is called an **angle**. The point where the two lines meet is called the **vertex** of the angle. Figure 11-2 represents an _____.

    angle

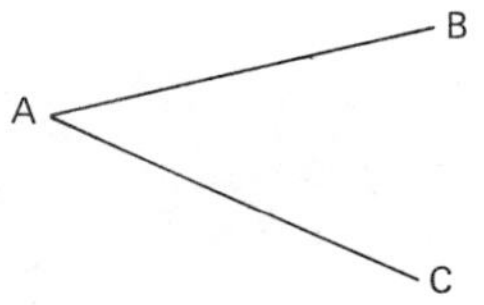

FIGURE 11-2

11. The point *A* in Fig. 11-2 is called the _____.

    vertex

12. The angle in Fig. 11-2 would be called ∢ *BAC*, where the symbol ∢ is read *angle*. How would you read ∢ *DEF*?

    angle *DEF*

13. A 90° angle is often called a **right angle**. If two lines meet to form an angle of 90°, the lines are said to be **perpendicular**. Are two lines perpendicular if they form a right angle?

    yes

14. If an angle is less than 90°, it is called **acute**; if it is greater than 90°, it is called **obtuse**. Is the angle in Fig. 11-3 acute or obtuse?

    acute

FIGURE 11-3

15.   Is the angle in Fig. 11-4 acute, obtuse, or right?

obtuse

FIGURE 11-4

16.   Is the angle in Fig. 11-5 acute, obtuse, or right?

right

FIGURE 11-5

17.   Two lines are **parallel** if they are perpendicular to the same line at different points on the line. We assume that two parallel lines will never meet. Are the lines *a* and *b* parallel in Fig. 11-6? (The symbol ⌐ means the angle is a right angle.)

Yes, since they are perpendicular to the same line.

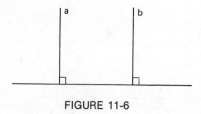

FIGURE 11-6

18.   Are lines *a* and *b* parallel in Fig. 11-7?

No, since they are *not* perpendicular to the same line. (Note that if you extend the lines, they will eventually meet or cross.)

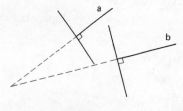

FIGURE 11-7

19.   A plane figure formed by two pairs of parallel lines that are perpendicular to each other is called a **rectangle**. Does the outline of this page represent a rectangle?

yes

20.   A rectangle in which the sides are all the same length is called a **square**. Does the outline of this paper form a square?

No, the sides are not all of equal length.

21. A triangle is a closed plane figure with three sides. It is called a **triangle** because the sides form three angles. Does Fig. 11-8 represent a triangle?

yes

FIGURE 11-8

22. A right triangle has an angle of 90°. Draw a right triangle.

FIGURE 11-9A

23. A **circle** is the figure formed by a closed plane curve, the points of which are all the same distance from a point within the circle called the *center*. In the circle in Fig. 11-10 are line segments *a* and *b* the same length?

Yes, since points on the circle are the same distance from the center.

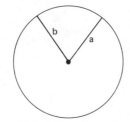

FIGURE 11-10

24. Lines that extend from the center of the circle to a point on the circle are called **radii**. Is line *a* in Fig. 11-10 a *radius* of the circle?

yes

25. A **diameter** of a circle is any line segment through the center of the circle with end points on the circle. Which of the line segments in Fig. 11-11 is a diameter, *AB* or *CD*?

AB, because it goes through the center.

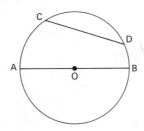

FIGURE 11-11

26. Refer to Fig. 11-11. The diameter *AB* of the circle is composed of two radii, *AO* and *OB*. Thus, the length of a diameter of a circle is twice the length of a radius or, equivalently, the length of a radius of a circle is _____ the length of a diameter.

one-half

27. A **trapezoid** is a four-sided plane figure in which *exactly* two of the sides are parallel. Is a rectangle a trapezoid?

No, since in a trapezoid two sides are parallel and the other two sides are *not* parallel, but in a rectangle both sets of opposite sides are parallel.

28. Solid figure definitions are similar to those for plane figures. Two planes are perpendicular if the angle formed by the planes is a right angle. If the plane surface of a wall and the plane surface of a floor form a right angle, they are _____.

perpendicular

29. A **rectangular solid** is a solid figure formed by perpendicular rectangles. Thus, a cardboard box would represent a _____ _____.

rectangular solid

30. A rectangular solid formed by squares is called a **cube**. Give an example of a cube.

Any rectangular solid with squares for sides, for example, ice cubes or a square-sided box.

31. A **sphere** is a solid, curved figure such that every point on the surface is the same distance from the point within the sphere called the *center*. A ball is an example of a _____.

sphere

Exercise Set 11-1.   Work all the problems before checking your answers.

1. A line has __no__ dimensions.
2. A geometric object of 0 dimension is called a _____.
3. A plane has __l & w__ dimensions.
4. A solid has __l w h__ dimensions.
5. The surface of a wall represents a __Plane__.
6. The surface of a (flat) penny represents a __boundries__.
7. A beach ball is an example of __Spere__.
8. There are __90__ degrees in a right angle.
9. The figure formed by two lines extending from the same point is called an __angle__.
10. The angle formed by a straight line is __180__ degrees.
11. Define acute and obtuse angles. __more than 90__

12. Two lines perpendicular to the same line at different points are said to be ___*parrell*___. _____

13. Draw a rectangle. ⬭ _____

14. A triangle has ___3___ sides. _____

15. Define the term *radius*. ___*From centu to oaste a a circle*___ _____

_____

4 s·sc 2·99 _____

16. Define the term *trapezoid*. _____

17. Give an example of a sphere. _____

18. Give an example of a rectangular solid. _____

11-2 AREA FORMULAS

Area refers to the amount of surface enclosed by a plane figure. In this section we shall look at formulas for finding the areas of several plane figures.

_____

32. When measuring the length of an object, you must start with some basic unit of measurement such as an inch, yard, mile, or meter. The same is true when determining the surface enclosed by a plane figure. To be consistent, the unit used will be based on the unit used in measuring the length of a side. Thus, if we have a square of side 1 unit in length, the area will be defined to be 1 square unit. A rectangle formed by 8 such square units would have an area of 8 square units. What is the area of a rectangle 1 unit wide and 8 units long?

8 square units

33. What is the area of a rectangle 2 units wide and 4 units long?

8 square units

34. The width and length of each of the rectangles above could be used to determine the area. For example, $1 \cdot 8 = 8$ and $2 \cdot 4 = 8$. The area of a rectangle is given by the formula $A = lw$, where A is the area, l is the length, and w is the width of the rectangle. The area of a rectangle of length 4 and width 3 is $A = 4 \cdot 3$ or 12. What is the area of a rectangle with length 5 and width 4?

20

35. The area of a square of side s units in length is $s \cdot s$ or s^2 since a square is a rectangle. Find the area of a square of side 7 inches.

49 square inches

36. Find the area of a square of side 2 feet.

4 square feet

37. Figure 11-12 is a right triangle. If we draw lines (dashed) parallel to
    the two sides forming the right angle, we shall form a rectangle.
    What is the area of the rectangle?   12

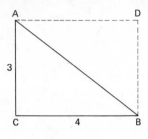

FIGURE 11-12

38. The triangles *ABC* and *ABD* in Fig. 11-12 have the same area. Thus,
    the area of triangle *ABC* is one-half the area of the rectangle
    or _____.   6

39. The area of a right triangle can thus be found by taking one-half
    times the product of the length and width of the corresponding
    rectangle. In the more general case, as in Fig. 11-13, the area is
    again one-half of the area of the rectangle formed. (Area of triangle
    ABD = area of triangle *ABC* + area of triangle *BCD*
      $= (\frac{1}{2}BC \cdot AC) + (\frac{1}{2}BC \cdot CD) = \frac{1}{2}BC \cdot (AC + CD) = \frac{1}{2}BC \cdot AD$.)
    The perpendicular line drawn from point *B* to line *AD* is called the
    height of triangle *ABD* where *AD* is the *base*. Thus, the formula for
    the area of a triangle is $A = \frac{1}{2}bh$, where *b* is the length of the base
    and *h* is the height. What is the height in Fig. 11-13?   4

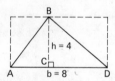

FIGURE 11-13

40. What is the base in Fig. 11-13?   8

41. What is the area of the triangle in Fig. 11-13? $A = \frac{1}{2}bh$
                                                                           $= \frac{1}{2}(8)(4)$
                                                                           $= 16$

42. What is the area of the triangle in Fig. 11-14? $A = \frac{1}{2}bh$
                                                                           $= \frac{1}{2}(7)(2)$
                                                                           $= 7$

FIGURE 11-14

43. The formula for the area of a circle is $A = \pi r^2$ where π (pronounced *pie*) is an irrational number approximately equal to 3.1416 and r is the radius of the circle. What is the radius in the circle in Fig. 11-15?

4

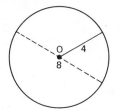

FIGURE 11-15

44. The area of the circle in frame 43 is

$$\pi(4)^2 = 3.1416(16) = 50.266.$$

What is the area of a circle of radius 2"?
(Since 3.1416 is an approximation for π, the answer should be rounded off to 5 significant digits using the rules of Section 2-2.)

$A = \pi r^2$
$\quad = 3.1416(2)^2$
$\quad = 3.1416(4)$
$\quad = 12.566$ square inches

45. In practice it is easier to find the diameter of the circle. The line segment AB is a diameter of the circle in Fig. 11-16. What is the length of the diameter?

4

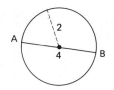

FIGURE 11-16

46. Since a diameter is composed of 2 radii, the diameter is twice the length of a radius or the length of a radius is $\frac{1}{2}$ of the diameter. Thus $r = \frac{1}{2}d$, where r is the radius and d is the diameter of a circle. If we replace r by $\frac{1}{2}d = d/2$ in the formula for the area of a circle, we find

$$A = \pi r^2 = \pi\left(\frac{d}{2}\right)^2 = \pi\left(\frac{d}{2}\right)\cdot\left(\frac{d}{2}\right) = \pi\frac{d^2}{4} = \frac{\pi d^2}{4}$$

The area of the circle in Fig. 11-16 is found using this formula.

$$A = \frac{\pi d^2}{4} = \frac{\pi(4)^2}{4} = \frac{(3.1416)(16)}{4} = 12.566$$

What is the area of a circle with diameter 6 inches?

$A = \dfrac{\pi d^2}{4}$

$\quad = \dfrac{3.1416(6)^2}{4}$

$\quad = \dfrac{3.1416(36)}{4}$

$\quad = 28.274$ in.^{2}

Exercise Set 11-2. Work all the problems before checking your answers. Find the area of each of the following:

1. A rectangle with a length of 13 and a width of 6. _____

2. A square of side 3″. _____

3. A triangle with a base of 12 and a height of 5. _____

4. A circle of diameter 14. _____

5. A circle of radius 9. _____

11-3 VOLUMES

Volume refers to the amount of space a solid object takes up. Again, we use a basic unit derived from the unit of length used, as we did in determining areas.

47. The basic unit in volume is a cube formed by squares of side 1 unit long. To find the volume of a rectangular solid, we multiply the length times the width to find the area of a side and multiply that by the height to obtain the volume. That is, $V = lwh$. Find the volume of a rectangular solid with the following dimensions:
length: 4 feet
width: 2 feet
height: 6 feet

$V = lwh$
$\quad = (4)(2)(6)$
$\quad = 48$ cubic feet

48. Find the volume of a rectangular solid of length 10 feet, width 2 yards, and height 7 feet (*Note*: 2 yards = 6 ft).

$V = lwh = (10)(6)(7)$
$\quad = 420$ cubic feet

49. The formula for the volume of a cube can be written as $V = S^3$, where S is the length of a side. Find the volume of a cube of side 6 inches.

$V = S^3 = 6^3 = 216$ cubic inches

50. The formula for the volume of a sphere is $V = \frac{4}{3}\pi r^3$, where r is the radius. Find the volume of a sphere with radius 4 inches.

$V = \frac{4}{3}\pi r^3$
$\quad = \frac{4}{3}(3.1416)(4)^3$
$\quad = \frac{4}{3}(3.1416)(64)$
$\quad \approx 268.08$ in.^{3}

51. A cylinder is a solid object in the general shape of a tin can. To find the volume of a cylinder you first find the area of the circle that is the base of the cylinder and multiply this by the height of the cylinder. Thus, the formula is $V = \pi r^2 h$ where r is the radius and h the height or $V = \pi d^2 h/4$, where d is the diameter of the circle and h is the

height. What is the volume of a cylinder with a radius of 3″ and a
height of 5″?

$$V = \pi r^2 h$$
$$= (3.1416)(3)^2(5)$$
$$= (3.1416)(9)(5)$$
$$= 141.37 \text{ cubic inches}$$

Exercise Set 11-3.   Work all the problems before checking your answers.
Find the volume of each of the following:

1.   A rectangular solid of length 4, width 3, and height 6.

2.   A cube of side 7″.

3.   A sphere with radius 3 inches.

4.   A cylinder of diameter 6″ and height 8″.

5.   A cylinder of radius 3 inches and height 8 inches.

11-4   THE PYTHAGOREAN THEOREM

One of the more important theorems of geometry deals with a property of right
triangles. It is called the *Pythagorean theorem* after the Greek mathematician
Pythagoras who is usually credited with formulation of the theorem. This theorem
has many practical applications, some of which you will see in this section.

---

52.   Ancient mathematicians knew that if the sides of a triangle were in
the ratio of 3 to 4 to 5, then the triangle was a right triangle. In fact,
they knew of other ratios that resulted in right triangles. They were
interested in these numbers because right triangles were so much
easier to work with. One of the properties of these numbers that was
discovered by this study was the Pythagorean theorem. It states that
the square of the side opposite the right angle equals the sum of the
squares of the other sides. Referring to Fig. 11-17, the theorem can be
stated as $c^2 = a^2 + b^2$. Thus, since the 3-4-5 triangle is a right
triangle, we know that $5^2 = 4^2 +$ _____.

3^2

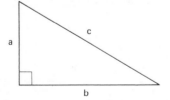

FIGURE 11-17

53.   Refer to Fig. 11-17. If we know $a = 5$ and $b = 12$, what is the length
of c?

To find the length of side c we can use the Pythagorean theorem. It
states that $c^2 = a^2 + b^2$ or $c^2 =$ _____ + _____.

5^2; 12^2

54.   Thus, $c^2 = 5^2 + 12^2$ ·
$$= 25 + 144$$
$$= 169$$

To find c, we need only take the square root of 169. Find c.

$$c = \sqrt{169}$$
$$= 13 \text{ (since } 13^2 = 169)$$

55. Find side b in the right triangle of Fig. 11-17 if $a = 3$ and $c = 5$.

$$c^2 = a^2 + b^2$$
$$5^2 = 3^2 + b^2$$
$$25 = 9 + b^2$$
$$16 = b^2$$
$$\sqrt{16} = b$$
$$4 = b$$

56. In the problem below, c is irrational since $\sqrt{5}$ cannot be simplified.

Example: Find c in a right triangle if $a = 1$ and $b = 2$.

$$c^2 = 1^2 + 2^2$$
$$= 1 + 4$$
$$= 5$$
$$c = \sqrt{5}$$

Find b in the right triangle if $a = 1$ and $c = 2$.

$$c^2 = a^2 + b^2$$
$$2^2 = 1^2 + b^2$$
$$4 = 1 + b^2$$
$$3 = b^2$$
$$\sqrt{3} = b$$

Exercise Set 11-4. Work all the problems before checking your answers.

1. State the Pythagorean theorem.

2. Find side b in a right triangle if $a = 6$ and $c = 10$.

3. Find the hypotenuse, c, in a right triangle if $a = 5$ and $b = 12$.

4. Find side a in a right triangle if $b = 7$ and $c = 11$.

SELF-TEST

1. State how many dimensions each object has.
    (a) circle
    (b) point
    (c) line
    (d) solid

2. How many degrees are there in a right angle?

3. Define the terms *acute* and *obtuse*.

4. Define *perpendicular lines*.

5. Draw a rectangular solid.

6. Find each of the following areas.
    (a) square of side 7 inches
    (b) circle of diameter 12 inches
    (c) triangle of height 3 inches and base 9 inches

7. Find each volume.
   (a) A rectangular solid of length 5, width 4, and height 10. _____
   (b) A sphere of radius 5". _____

8. Find side b of a right triangle if $a = 10$ and $c = 20$. _____

APPLIED PROBLEMS

A. The compression ratio, C, of an automobile is defined as the ratio of the total volume, V_T, of the cylinder to the volume in the cylinder when the piston is at the top of the stroke (or top dead center), V_{TDC}. Also, the total volume equals the volume at top dead center plus the piston displacement, $P.D.$, or $V_T = V_{TDC} + P.D.$ Thus, compression ratio $= (V_{TDC} + P.D.)/V_{TDC}$. Find the missing values in the table below using the concepts of this chapter. (d is the bore of the cylinder and h the length of the stroke.) Piston displacement is defined in Problem O.

	V_{TDC}	d	$P.D.$	V_T	C	h
1.	7.8 in.³	4 in.				5 in.
2.	7.2 in.³	3.875 in.				4.5 in.
3.	6.8 in.³	3.625 in.				4.25 in.
4.	7.72 in.³	4.025 in.				4.3 in.

B. A sector is the portion of a circle between two radii of a circle. The area of a sector is directly proportional to the angle, θ, between the radii. Thus $A_S/A_C = \theta/360°$, where A_S is the area of the sector and A_C is the area of the circle. Find the missing values in the table below. (r is the radius of the circle.)

	r	θ	A_C	A_S
1.	3	45°		
2.	4			3.1416
3.	6	35°		
4.	12			37.699
5.	10	100°		

Example: 1. $A_C = \pi r^2 = 28.27$

Thus,

$$\frac{A_S}{28.27} = \frac{45}{360}$$

$$A_S = (45)(28.27) \div 360 = 3.534$$

C. The circumference, *C*, ground area covered, *A*, and volume, *V*, of a circular tank can be found if you know the diameter, *D*, of the tank and its height, *H*. Find the missing values in the table below. (*Note*: $C = \pi D$.)

	D	H	C	A	V
1.	18	12			
2.		15	69		
3.			94		12,600
4.	25	10			
5.		20	62.8		

D. The reach of an *X*-foot ladder is *Y* feet when the foot of the ladder is $\frac{1}{4}$ of *X* (*X*/4) feet from the building wall. Find the missing values in the table below using the Pythagorean theorem.

	X	X/4	Y
1.	72		
2.		12	
3.	60		
4.		8	
5.	36		

E. The safe speed of an abrasive wheel is *X* feet per minute. This is equivalent to *Y* rpms on a wheel with a *Z*-inch diameter. Find the missing values in the table below.

	X	Y	Z
1.	5,000		14
2.		1,000	15
3.	5,280		16
4.	5,000	1,200	
5.		1,600	10
6.	4,800		8

Example: 1. The circumference of the wheel is $\pi(14'') \approx 44''$.

$$44'' = \tfrac{44}{12} = 3\tfrac{2}{3} \text{ feet}$$

Thus, in one revolution the wheel turns $3\tfrac{2}{3}$ feet. The rpms are therefore $5,000 \div 3\tfrac{2}{3} \approx 1,364$.

F. X-inch diameter round stock or larger must be used to mill a square Y inches on a side. Find the missing values in the table below. (*Hint*: Draw a diagram and use the Pythagorean theorem.)

	X	Y
1.	2	
2.		2
3.	3	
4.		$2\tfrac{1}{2}$
5.	1	

G. The area, A, of a ring can be found using the inside diameter, X, and the outside diameter, Y. Find the missing values in the table below.

	X	Y	A
1.	4″	8″	
2.	6″	9″	
3.	7″	8″	
4.	10″		75.4
5.		13″	82.46
6.	6″		50.27

Example: 1. The area of the ring is the area of the large (8″ diameter) circle minus the area of the small (4″ diameter) circle.

$$\frac{\pi(8'')^2}{4} \approx 50.3 \text{ in.}^2 \quad \text{and} \quad \frac{\pi(4'')^2}{4} \approx 12.6 \text{ in.}^2$$

$$A = 50.3 \text{ in.}^2 - 12.6 \text{ in.}^2 = 37.7 \text{ in.}^2$$

H. A hexagonal nut is milled from round stock with a radius of X inches. How deep should the cut, Y, be in each case? See Fig. 11-18.

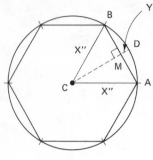

FIGURE 11-18

	X	Y
1.	$1''$	
2.	$1\frac{1}{2}''$	
3.	$\frac{3}{4}''$	
4.	$1\frac{1}{4}''$	
5.	$2''$	

Example: 1. The radii, AC and BC, bisect the interior angles, which are $120°$ (see Chapter 3, Problem D). Thus, $\angle CBA = \angle CAB = 60°$. Thus, $\angle BCA = 60°$ (why?) and triangle ABC is an equilateral triangle (i.e., all sides are the same length), so $AB = 1''$. $BM = MA$ since triangles BCM and ACM are congruent (why?). Thus, $BM = \frac{1}{2}(1'') = \frac{1}{2}''$. Using the Pythagorean theorem, $BC^2 = CM^2 + BM^2$ or $(1'')^2 = CM^2 + (\frac{1}{2}'')^2$. Thus, $1 = CM^2 + \frac{1}{4}$ or $\frac{3}{4} = CM^2$, so $.75 = CM^2$, $\sqrt{.75} = CM$, or $CM \approx .866''$.

$$Y = CD - CM = 1'' - .866'' = .134''$$

I. One method of determining the flow of a fluid in a pipeline sometimes requires determining the cross-sectional area filled by water (see the shaded portion in Fig. 11-19). Since in general the depth of the water, EF, can readily be determined (and the diameter of the pipe, DC, is known), but not AB, the Pythagorean theorem is used to determine the length of AB. Then geometry can be used to approximate the area if we use the trapezoid $ABCD$ to represent the area of the top-shaded portion. Find the area below given the following data:

	EF	*DC*	*Area*
1.	$4''$	$5''$	
2.	10 cm	12 cm	
3.	.5 m	.8 m	
4.	$1\frac{1}{4}''$	$1\frac{1}{2}''$	
5.	$2''$	$2\frac{1}{2}''$	
6.	13.7 cm	16.2 cm	

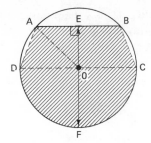

FIGURE 11-19

Hints: *AO* and *OF* are radii and thus one-half the diameter, *DC*.

$$EO = EF - OF$$
$$AE = EB$$

Use a combination of areas.

J. In Problem I, if the depth, *EF*, is less than half the diameter, a different diagram results (see Fig. 11-20). Find the area of the shaded portion of this diagram for the data below:

	EF	*DC*	*Area*
1.	1″	3″	
2.	6 cm	15 cm	
3.	$1\frac{3}{4}''$	5″	
4.	6.4 cm	16 cm	
5.	5.1″	12.3″	

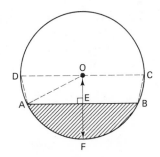

FIGURE 11-20

K. In order to check his work with screw threads, a machinist may want to measure the pitch diameter. Since this cannot be done directly, he must use accurately machined wires and "mike" (use a micrometer to measure) across the wires. Some calculation must be performed since this measurement is larger than the pitch diameter by 2*S* (see Fig. 11-21).

If the diameter of the wire is X inches and the distance AC is $\frac{1}{4}$ the pitch, P, find S using the Pythagorean theorem for each problem below.

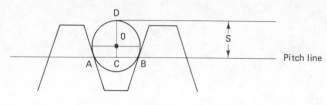

FIGURE 11-21

	X	P	S
1.	.0288	.05"	
2.	.155	.2"	
3.	.0361	.0625"	
4.	.0444	.077"	
5.	.0525	.091"	

Hint: $S = OD + OC$

    OD and OA are radii of the circle.

L. Find the areas of the following figures by breaking up the figure into a combination of polygons for which you know a formula for the area. The first figure has been broken up into two triangles and a rectangle by the dotted lines to illustrate the technique.

1.

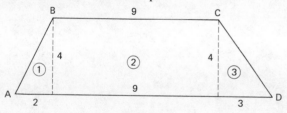

FIGURE 11-22

2.

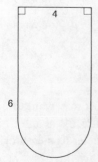

FIGURE 11-23

3.

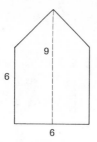

FIGURE 11-24

4.

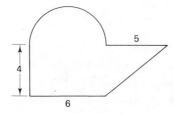

FIGURE 11-25

M. Find the volumes of the following figures by breaking them up into smaller figures for which you know volume formulas. The first one has been broken up for you.

1.

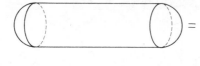

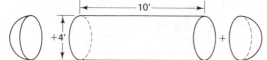

FIGURE 11-26

2.

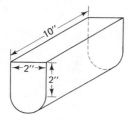

FIGURE 11-27

N. The impedance, Z, resistance R, and reactance, X, of a circuit such as the one shown below are related by the formula $Z^2 = R^2 + X^2$. Find the missing values in the table.

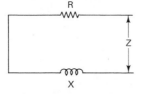

FIGURE 11-28

	Z	R	X
1.		30	40
2.	48		42
3.	50	23	
4.		35	40
5.	14		10

O. Refer to Figure 11-29. The *piston displacement* of one cylinder is the volume of the cylinder displaced when the piston moves from the bottom of the stroke to the top of the stroke. This volume is illustrated by the shaded area in Figure 11-29(b). Find the volume, *V*, for each instance in the table below if *d* is the bore and *h* is the length of the stroke.

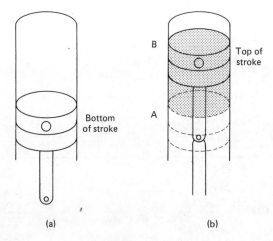

FIGURE 11-29

	d	h	V
1.	$3\frac{3}{8}''$	5.2″	
2.	9 cm	10 cm	
3.	4″	$4\frac{1}{8}''$	
4.	9.2 cm	9.4 cm	
5.	$3\frac{7}{8}''$	$3\frac{7}{8}''$	

12

applied trigonometry

OBJECTIVES

After completing this chapter the student should be able to
1. Define the sine, cosine, tangent, secant, cosecant, and cotangent for right triangles.
2. Find the sine, cosine, tangent, secant, cosecant, or cotangent of a specified angle of a right triangle.
3. Use the trigonometric functions to solve a right triangle.
4. Use a table to find the sine, cosine, or tangent of an angle between 0° and 180°.
5. Use a table to find an angle, given its sine, cosine, or tangent.
6. State the law of sines and the law of cosines.
7. Use the law of sines and the law of cosines to solve given triangles.

(1–14)   **1.**   (a) Define sine, tangent, and secant for right triangles.

_____
_____
_____

   (b) Find cos A, cot B, csc A in Fig. 12-1:

cos A _____
cot B _____
csc A _____

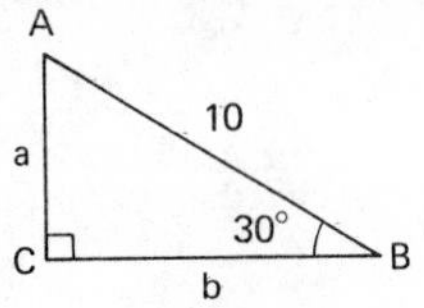

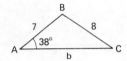

FIGURE 12-1

   (c) If X and Y are complementary angles and tan $X = \frac{3}{4}$, then
        cot $Y =$ _____.

_____

(15–38)   **2.**   Solve the right triangle in Problem 1(b).

$a =$ _____
$b =$ _____
$\angle A =$ _____

(39–52)   **3.**   (a) Find sin 125°.
          (b) Find cos 135°.
          (c) Solve the triangle in Fig. 12-2.

_____

$\angle B =$ _____
$\angle C =$ _____
$b =$ _____

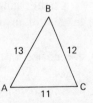

FIGURE 12-2

(53–61)   **4.**   (a) Find the measure of angle A from the trigonometric table
               in the back of the book if cos $A = {}^{-}.90631$.
          (b) Solve the triangle in Fig. 12-3.

$\angle A =$ _____
$\angle B =$ _____
$\angle C =$ _____

B

13   12

A   11   C

FIGURE 12-3

12-1   DEFINITIONS

The sides of a right triangle and the angles opposite are related. A study of this
relation leads to the trigonometry functions that will be defined in this section.

1. A basic proposition of geometry states that the sum of the angles of a triangle is 180°. Thus, if we know two of the angles, we can find the third. How many degrees are in angle A in Fig. 12-4?

    angle A + angle B + angle C = 180°
    angle A + 35° + 90° = 180°
    angle A = 180° − 125° = 55°

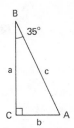

FIGURE 12-4

2. The side opposite the right angle in a right triangle is called the **hypotenuse**. For convenience we shall call the side other than the hypotenuse, that forms one edge of an angle, the adjacent side of the angle. Thus, in Fig. 12-4 side _____ is adjacent to angle B.

    a

3. Which side is adjacent to angle A in Fig. 12-4?

    side b

4. Angle B in Fig. 12-5 can be determined by the sides of the triangle. We define three trigonometric functions.

    > The **sine** (abbreviated sin) of an angle is defined to be the ratio of the opposite side over the hypotenuse.

    In Fig. 12-5 the sin A = _____.

    $\dfrac{\text{opposite}}{\text{hypotenuse}} = \dfrac{a}{c}$

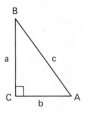

FIGURE 12-5

5. > The **cosine** (abbreviated cos) of an angle is the ratio of the adjacent side to the hypotenuse.

    Thus in Fig. 12-5 the cos A = _____.

    $\dfrac{\text{adjacent}}{\text{hypotenuse}} = \dfrac{b}{c}$

6. > The **tangent** (abbreviated tan) of an angle is the ratio of the opposite side to the adjacent side.

    Thus in Fig. 12-5 the tan A = _____.

    $\dfrac{\text{opposite}}{\text{adjacent}} = \dfrac{a}{b}$

7. **Example:** Find the tan B in Fig. 12-6.

$$\tan B = \frac{\text{opposite}}{\text{adjacent}} = \frac{4}{3}$$

Find the tan A in Fig. 12-6.

$$\tan A = \frac{\text{opposite}}{\text{adjacent}} = \frac{3}{4}$$

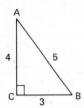

FIGURE 12-6

8. Find the cos B in the Fig. 12-6.

$$\cos B = \frac{\text{adjacent}}{\text{hypotenuse}} = \frac{3}{5}$$

9. Find the sin B in Fig. 12-6.

$$\sin B = \frac{\text{opposite}}{\text{hypotenuse}} = \frac{4}{5}$$

10. The **sine** of an angle is defined as the ratio of the opposite side to the hypotenuse of the right triangle, or for some angle A,

$$\sin A = \frac{\text{opposite side}}{\text{hypotenuse}}$$

If we look at the reciprocal of the ratio on the right, we obtain a new function. This function is called the **cosecant** (abbreviated csc).

> The **cosecant** of an angle is defined as the ratio of the hypotenuse to the opposite side.

Note that the cosecant is the reciprocal of the sine so that

$$\csc A = \frac{1}{(\underline{\quad\quad})}.$$

$\sin A$

11. If we look at the reciprocals of the cosine and tangent, we obtain two more functions called the *secant* and *cotangent*, respectively. Their definitions are stated below. The abbreviation for secant is sec and the abbreviation for cotangent is cot.

> $$\text{secant } A = \frac{\text{hypotenuse}}{\text{adjacent side}}$$
>
> $$\text{cotangent } A = \frac{\text{adjacent side}}{\text{opposite side}}$$

Note that the cot $A = 1/\tan A$ and sec $A = $ _____.

$$\frac{1}{\cos A}$$

12. Find the sin A, tan A, sec A, cos B, cot B, csc B for the triangle in Fig. 12-7.

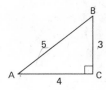

FIGURE 12-7

sin $A = \frac{3}{5}$

tan $A = \frac{3}{4}$

sec $A = \frac{5}{4}$

cos $B = \frac{3}{5}$

cot $B = \frac{3}{4}$

csc $B = \frac{5}{4}$

13. Note the answers in frame 12. The sin $A = $ cos B, tan $A = $ cot B, sec $A = $ csc B. It is always true for complementary angles (two angles whose sum is 90°) that the sine, tangent, or secant of an angle is equal to the cosine, cotangent, or cosecant, respectively, of the complementary angle. Thus, if E and F are complementary angles and sin $E = \frac{1}{3}$, then cos $F = $ _____.

$\frac{1}{3}$

14. If X and Y are complementary angles and tan $X = \frac{5}{2}$, what is cot Y?

cot $Y = \frac{5}{2}$

Exercise Set 12-1.   Work all the problems before checking your answers.

1. Define the sine, cosine, tangent, secant, cosecant, and cotangent.

_____

_____

_____

_____

_____

_____

2. Find the sin A, tan B, and sec B in Fig. 12-8.

FIGURE 12-8

_____

_____

_____

3. Find the csc A, cos B, and cot B in Fig. 12-9.

FIGURE 12-9

_____

_____

_____

4. If R and S are complementary angles and sec $R = \frac{5}{7}$, then csc $S = $ _____.

_____

12-2 SOLVING RIGHT TRIANGLES

Trigonometric functions can be used to find the sides of a triangle or the angles that are missing given enough information. For this reason extensive tables have been developed to aid in solving triangles through the use of "trig" functions. In this section we shall look at some examples of how trig functions can be used.

15. If we know that the sine of an angle is equal to the ratio $\frac{4}{5}$, then we know that the ratio of the side opposite to the hypotenuse must be $\frac{4}{5}$. If we know that the hypotenuse is 20, then we can find the side opposite the angle through the use of a proportion. Let $\sin A = \frac{4}{5}$ in Fig. 12-10. Then by definition $\sin A = a/20$. Substituting $\frac{4}{5}$ for $\sin A$, we obtain the proportion $\frac{4}{5} = a/20$. Solve this proportion for a.

$$(100)\frac{4}{5} = (100)\frac{a}{20}$$
$$80 = 5a$$
$$16 = a$$

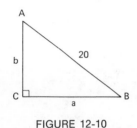

FIGURE 12-10

16. If the $\cos A$ in Fig. 12-11 is $\frac{1}{2}$, find the hypotenuse, c.

$$\cos A = \frac{11}{c}$$
$$\frac{1}{2} = \frac{11}{c}$$
$$c = 22$$

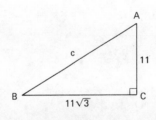

FIGURE 12-11

17. Refer to the Natural Trigonometric Table. To find the sine, tangent, cotangent or cosine of an angle less than 45° using this table, the following procedure is used.

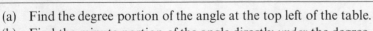

> (a) Find the degree portion of the angle at the top left of the table.
> (b) Find the minute portion of the angle directly *under* the degree portion of the angle found in step (a). (It is under the heading, '.)
> (c) Find the desired trigonometric function at the *top* of the table. Read the decimal numeral under the correct heading (Sin, Tan, Ctn, or Cos) to the right of the minute portion.

Example: Find the sin 30°10′.
   (a)   30°, the degree portion of 30°10′, is found at the top left of
         page 271 in the table.
   (b)   Going down the far left column, find the 10, the minute
         portion of 30°10′.
   (c)   Find the heading Sin (the second column). The decimal
         numeral in this column to the right of the 10 is .50252. Thus,
         sin 30°10′ = .50252 (rounded off to 5 digits). Find sin 31°5′.   .51628

18. Use the table to find cos 35°. (*Hint:* 35° = 35°0′.)   .81915

19. Find cos 40°.   .76604

20. Find tan 38°.   .78129

21. Find tan 15°.   .26795

22. Find sin 25°15′.   .42657

23. Find cos 0°30′.   .99996

24. Find tan 41°45′.   .89253

25. Find cos 40°15′.   .76323

26. Find tan 4°.   .06993

27. To find the sine, tangent, cotangent or cosine of an angle between 45°
     and 90° the following procedure is used.

   (a)   Find the degree portion of the angle at the *bottom right* of the
         table.
   (b)   Find the minute portion of the angle in the column directly
         above the degree portion found in step (a). (This column has
         a ′ at the bottom of it.)
   (c)   Find the correct trigonometric function at the bottom of the
         table. Read the decimal numeral in this column to the left of
         the minute portion.

Example: Find tan 57°11′.
   (a)   57° is found at the bottom of page 272 in the table.
   (b)   The 11 is then found directly above the 57°.
   (c)   The third column from the left has Tan at the bottom. The
         decimal numeral in this column to the left of the 11 is 1.5507.

Thus, tan 57°11′ = 1.5507 (rounded off to 5 digits). Find cos 47°33′.   .67495

28. Find cos 60°.   .50000

29. Find tan 60°.   1.7321

30. Find tan 72°15′.   3.1240

31. Find sin 75°30′.   .96815

32. Find cos 85°29′.   .07875

33. Find tan 80°.   5.6713

34. You now have the tools you need to solve right triangles. For example, in the right triangle in Fig. 12-12, we can find each side and angle from what is given. To find side *a*, we could use the sin *A* = sin 30° since the sine is the opposite side over the hypotenuse. Thus, sin 30° = *a*/2. Since sin 30° = .50000, we have .5 = *a*/2 and (2).5 = (2)*a*/2 or 1 = *a*. Thus, side *a* has length 1. Use the cos 30° to find side *b*.

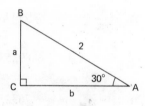

FIGURE 12-12

$$\cos 30° = \frac{\text{adjacent}}{\text{hypotenuse}} = \frac{b}{2}$$

$$.86603 = \frac{b}{2}$$

$$2(.86603) = b$$

$$1.7321 = b$$

(Answers are rounded off to 5 significant digits in accordance with the conventions of Section 2-2 since the table values are only accurate to 5 digits.)

35. Since the sum of the angles of a triangle is 180°, angle *B* in Fig. 12-12 must be 60°. We have thus solved the triangle. Solve the triangle in Fig. 12-13 by using the sine and tangent of the given angle.

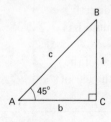

FIGURE 12-13

$$\sin 45° = \frac{1}{c}$$

$$.70711 = \frac{1}{c}$$

$$.70711c = 1$$

$$c = \frac{1}{.70711} = 1.4142$$

$$\tan 45° = \frac{1}{b}$$

$$1 = \frac{1}{b}$$

$$b = 1$$

$$\angle B + 45° + 90° = 180°$$

$$\angle B = 45°$$

36. If you are given two sides of a right triangle, you can solve the triangle also. For example, in the triangle in Fig. 12-14 we can find angle A by using the sine.

$$\sin A = \frac{\text{opposite}}{\text{hypotenuse}} = \frac{3}{5} = .6$$

Looking up .6 under the sine heading we see that .6 is closest to $36°52'$ in the table. Thus, $A = 36°52'$. If you want a closer approximation, you would have to use interpolation or a better table. Find angle B using the cosine.

$\cos B = \frac{3}{5} = .6$
$\measuredangle B = 53°8'$

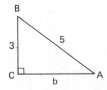

FIGURE 12-14

37. To find side b in Fig. 12-14 we could use the tangent or cosine of angle A. Find side b.

$\cos 36°52' = .80003$

$$.80003 = \frac{b}{5}$$

$4.0002 = b$ (Actually $b = 4$, but the angle was only approximate.)

38. To solve a right triangle:
   (a) Find a trigonometric function which involves two known values and one unknown value.
   (b) Write an equation involving this function, the unknown value, and the two known values.
   (c) Solve for the unknown using the trigonometric table.
   (d) Continue in this way until you have found all the unknowns.

Solve the right triangle in Fig. 12-15.

(a), (b) $\sin A = \frac{2}{7}$
(c) $\measuredangle A = 16°36'$
(d) $\measuredangle B = 73°24'$
    $\cos A = b/7$
    $.95832 = b/7$
    $7(.95832) = b$
    $6.7082 = b$

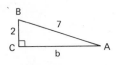

FIGURE 12-15

Exercise Set 12-2. Work all the problems before checking your answers.

1. Solve the right triangle in Fig. 12-16.

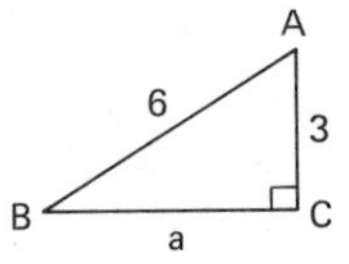

FIGURE 12-16

2. Solve the right triangle in Fig. 12-17.

FIGURE 12-17

12-3 THE LAW OF SINES

In this section we shall derive one of two laws which will be quite useful in dealing with triangles which do not have a right angle in them. Since most triangles are not right triangles, these two laws—the law of sines and the law of cosines—have many applications in technical fields.

39. Consider an arbitrary triangle *ABC*. We would like to look first at the relationship between the sines of the angles and the sides opposite the angles. Recall that the largest side of a triangle is opposite the larger angle. Also, if you refer to the trigonometric table, you will notice that as the angle increases, the sine increases. Thus, it seems plausible that the ratio of sines is equal to the ratio of the opposite sides. Consider angles *A* and *B* in Fig. 12-18, which shows two possible triangles. What we would like to prove is that

$$\frac{\sin A}{\sin B} = \underline{\hspace{4cm}}$$

$\dfrac{a}{b}$, the ratio of the opposite sides.

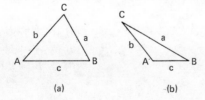

FIGURE 12-18

40. Before we can prove our assertion, we must come up with a more general definition of the sine of an angle since the one we have been using is only valid for angles less than or equal to 90°. Thus, we could not deal with the sine of angle *A* in Fig. 12-18(b) without a better definition. To do this we shall have to introduce the concept of a rectangular coordinate system. A rectangular coordinate system is formed by two number lines perpendicular to each other as shown in Fig. 12-19. The horizontal number line is called the *x*-axis and the vertical one, the *y*-axis.

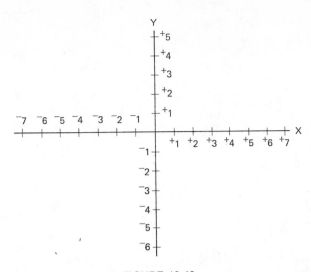

FIGURE 12-19

The point where the two axes cross is called the *origin*. In defining
the trigonometric functions the origin is always the vertex of the
angle. We also use a circle of arbitrary radius, *r*, with center at the
origin. With the radius along the positive *x*-axis as one of the radii,
we can form any angle. If we drop a perpendicular from the point
where the terminal side of the angle meets the circle, we are ready to
define the trigonometric functions. See Fig. 12-20. The Greek letter θ
refers to the angle being considered. The definitions are

$$\sin \theta = \frac{y}{r}$$

$$\cos \theta = \frac{x}{r}$$

$$\tan \theta = \frac{y}{x}$$

Notice that since *y* and *r* are positive and *x* is negative when θ is
between 90° and 180°, sin θ is positive, cos θ is negative, and tan θ is
_____ between 90° and 180°.

Negative, since y/x is a positive
divided by a negative.

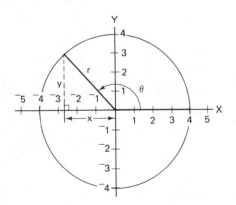

FIGURE 12-20

41. Consider 150°. In Fig. 12-21, compare sin 30° with sin 150° using the new definition. (Note that since triangles OAB and OCD are congruent, $y_1 = y_2$.)

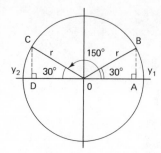

FIGURE 12-21

$$\sin 150° = \frac{y_2}{r}$$

$$\sin 30° = \frac{y_1}{r}$$

But since $y_1 = y_2$,
$\sin 150° = \sin 30°$.

42. (See Fig. 12-22.) In the general case, if we subtract the angle θ from 180° to find the angle $180° - \theta$, what can you say about $\sin \theta$ and $\sin (180° - \theta)$?

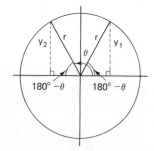

FIGURE 12-22

$\sin \theta = \sin (180° - \theta)$, since
$y_1 = y_2$. (Thus, if θ is between 90° and 180°, $\sin \theta = \sin (180° - \theta)$.)

43. Since $\sin \theta = \sin (180° - \theta)$ (θ between 90° and 180°), we can use the table on p. 255 and the procedure outlined in Section 12-2 to find the sine of an angle between 90° and 180°. What is sin 140°?

$$\sin 140° = \sin (180° - 140°)$$
$$= \sin 40°$$
$$= .64279$$

44. Find sin 120°.

$$\sin 120° = (180° - 120°)$$
$$= \sin 60°$$
$$= .86603$$

45. See Fig. 12-23. Note again that triangles OAB and OCD are congruent but that x_1 is positive and x_2 is negative. Compare $\cos \theta$ with $\cos (180° - \theta)$.

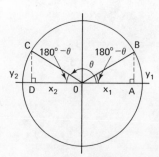

FIGURE 12-23

$$\cos \theta = \frac{x_2}{r}$$

$$\cos (180° - \theta) = \frac{x_1}{r}$$

Since $x_2 = -x_1$,
$\cos \theta = -\cos (180° - \theta)$

46. Refer to Fig. 12-23. Compare $\tan \theta$ and $\tan (180° - \theta)$.

$$\tan \theta = \frac{y_2}{x_2}$$

$$\tan (180° - \theta) = \frac{y_1}{x_1}$$

Since $y_2 = y_1$ and $x_2 = -x_1$,
we have $\tan \theta = -\tan (180° - \theta)$.

47. For angles between $90°$ and $180°$, we have

$$\sin \theta = \sin (180° - \theta)$$
$$\cos \theta = -\cos (180° - \theta)$$
$$\tan \theta = -\tan (180° - \theta)$$

For the other three functions we need only note that reciprocals have the same sign to see that $\csc \theta = \csc (180° - \theta)$, $\sec \theta = -\sec (180° - \theta)$ and $\cot \theta = -\cot (180° - \theta)$. Find $\cos 100°$.

$$\cos 100° = -\cos (180° - 100°)$$
$$= -\cos 80°$$
$$= ^{-}.17365$$

48. Find $\tan 135°$.

$$\tan 135° = -\tan (180° - 135°)$$
$$= -\tan 45°$$
$$= ^{-}1$$

49. We are now ready to prove that

$$\frac{\sin A}{\sin B} = \frac{a}{b}$$

In Fig. 12-24, if we drop a perpendicular from the vertex C to the extended line segment AB, we can prove our assertion. Using the fact that in Fig. 12-24(b) angle $CAD = 180° - \theta$ and that $\sin \theta = \sin (180° - \theta)$ we have that $\sin A = h/b$ and $\sin B = h/a$. Dividing, we find

$$\frac{a}{b}$$

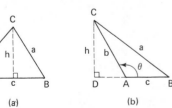

(a)   (b)

FIGURE 12-24

50.
$$\frac{\sin A}{\sin B} = \frac{a}{b}$$

can be written as

$$\frac{a}{\sin A} = \frac{b}{\sin B}$$

In the same way we can show this law holds for the other possible combinations, also. Thus, we have for any triangle

Law of Sines
$$\dfrac{a}{\sin A} = \dfrac{b}{\sin B} = \dfrac{c}{\sin C}$$

We can now use the law of sines to solve triangles that are not right triangles if we know two sides and the angle opposite one of them or two angles and the side opposite one of them.

Example: Solve the triangle in Fig. 12-25.

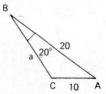

FIGURE 12-25

Since we know two sides and the angle opposite the one side, we can use the law of sines. Since $b = 10$, $c = 20$, and angle $B = 20°$, we have $\left(\text{using } \dfrac{b}{\sin B} = \dfrac{c}{\sin C}\right)$

$$\frac{10}{\sin 20°} = \frac{20}{\sin C}$$

Solving this proportion we find $10(\sin C) = 20(\sin 20°)$, or solving for $\sin C$,

$$\sin C = \frac{20(\sin 20°)}{10}$$
$$= 2(\sin 20°)$$
$$= 2(.34202)$$
$$= .68404$$

From the trigonometry table we see that $\sin 43° \ 10' = .68404$. From Fig. 12-25, however, it is obvious that angle C is larger than $43°10'$. Recalling that $\sin (180° - \theta) = \sin \theta$, we know that $\sin (180° - 43°10') = \sin 43°10'$. Thus, $\sin (180° - 43°10') = \sin$

$136°50' = .68404*$, also. Therefore, angle $C = 136°50'$. Since
angle $B = 20°$ and angle $C = 136°50'$, we know that angle
$A =$ _____ because the sum of the angles of a
triangle is $180°$.

$23°10'$
$136°50' + 20° = 156°50'$
$180° - 156°50' = 23°10'·$

51. From frame 50 we know all but the length of side a in the triangle
given. We again use the law of sines to find the length of side a.
The law of sines gives us the following proportion:

$$\frac{a}{\sin 23°10'} = \frac{10}{\sin 20°}$$

Solve the proportion for a.

Since $\sin 23°10' = .39341$ and
$\sin 20° = .34202$, we have

$$\frac{a}{.39341} = \frac{10}{.34202}$$

$$.34202a = 3.9341$$

$$a = \frac{3.9341}{.34202}$$

$$\approx 11.503$$

52. Use the law of sines to solve the triangle in Fig. 12-26.

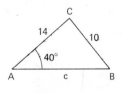

FIGURE 12-26

$$\frac{14}{\sin B} = \frac{10}{\sin 40°}$$

$$\frac{14}{\sin B} = \frac{10}{.64279}$$

$10 \sin B = 14(.64279)$
$10 \sin B = 8.99906$
$\sin B = .899906 \approx .89991$
$B = 64°9'$ (since it is acute)
Thus, $C = 75°51'$

$$\frac{c}{\sin 75°51'} = \frac{10}{\sin 40°}$$

$$\frac{c}{.96966} = \frac{10}{.64279}$$

$$.64279c = 9.6966$$

$$c = \frac{9.6966}{.64279}$$

$$\approx 15.085$$

---

* In using the law of sines, notice that quite often there are two possible solutions
for the angle. Without a picture we could not know which angle to choose and both
possibilities would have to be listed.

Exercise Set 12-3. Work all the problems before checking your answers.

1. $\sin(180° - \theta) = \sin \underline{\hspace{1.5cm}}$ _____

2. Find $\sin 137°$.   _____

3. Find $\cos 120°$.   _____

4. Find $\tan 98°$.   _____

5. Use the law of sines to solve the triangle in Fig. 12-27.

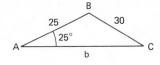

FIGURE 12-27

$\angle C = $ _____

$\angle B = $ _____

side $b = $ _____

12-4 THE LAW OF COSINES

We could not use the law of sines to solve a triangle in which you do not know the length of the side opposite the only given angle. In such a case we need a different law to solve the triangle. This is the *law of cosines*.

53. The Pythagorean Theorem only holds for right triangles. However, it can be shown that a slightly altered form of this famous theorem is true for all triangles. This is the law of cosines. The three forms of the law of cosines are written below. Three forms are needed, since if you only know one angle of the triangle you must use that form of the theorem which contains that angle. Which form would you use if you knew sides a and c and angle B?

$$b^2 = a^2 + c^2 - 2ac \cos B$$

Law of Cosines
$$c^2 = a^2 + b^2 - 2ab \cos C$$
$$b^2 = a^2 + c^2 - 2ac \cos B$$
$$a^2 = b^2 + c^2 - 2bc \cos A$$

---

54. We are now ready to use the law of cosines to solve some triangles. Consider the triangle in Fig. 12-28. Since we do not know any of the angles, we cannot use the law of sines. But we can use the law of cosines. In Fig. 12-28

$a = $ _____ $a = 13$

$b = $ _____ $b = 18$

$c = $ _____ $c = 11$

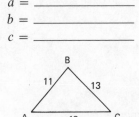

FIGURE 12-28

55. Substituting *a*, *b*, and *c* into $a^2 = b^2 + c^2 - 2bc \cos A$ we can find the measure of angle *A*.
$(13)^2 = (18)^2 + (11)^2 - 2(18)(11) \cos A$
$169 = 324 + 121 - 396 \cos A$
$169 = 445 - 396 \cos A$
$-276 = -396 \cos A$
$.69697 = \cos A$
Using the trig table, find the measure of angle *A*.

angle $A = 45°49'$

56. Refer to Fig. 12-28. If we want to find the measure of angle *B*, using the law of cosines, we would have to use the formula $b^2 = a^2 + c^2 - 2ac \cos B$. Use the law of cosines to find the measure of angle *B*. [*Hint*: Recall from frame 47 that $\cos \theta = -\cos(180° - \theta)$. Thus, if the value of the cosine of an angle of a triangle is negative, the angle is between 90° and 180° and can be found by subtracting the table value from 180°.]

$b^2 = a^2 + c^2 - 2ac \cos B$
$(18)^2 = (13)^2 + (11)^2$
$\qquad - 2(13)(11) \cos B$
$324 = 169 + 121 - 286 \cos B$
$34 = {}^-286 \cos B$
${}^-.11888 = \cos B$
From the trig table we find $83°10'$.
Thus, angle $B = 180° - 83°10'$.
$\quad 179°60'$
$- \quad 83°10'$
$\overline{\qquad 96°50'}$
angle $B = 96°50'$

57. Refer to Fig. 12-28. Since angle $A = 45°49'$ and angle $B = 96°50'$, what is the measure of angle *C*?

angle $C = 180° - 45°49' - 96°50'$
$= 37°21'$

58. Use the law of cosines to solve the triangle in Fig. 12-29 as follows: Use the law $b^2 = a^2 + c^2 - 2ac \cos B$ to find *b*. (*Note*: This formula is used because we know angle *B*.)
$b^2 = (10)^2 + (13)^2 - 2(10)(13) \cos 49°$
$b^2 = 100 + 169 - 260(.65606)$
$b^2 = 269 - 170.5756$
$b^2 = 98.4244$
$b \approx$ _____ (to 5 significant digits)

$b \approx \sqrt{98.4244} \approx 9.9209$
(*Note*: Use of a calculator is suggested for finding square roots.)

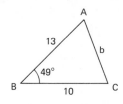

FIGURE 12-29

59. Refer to Fig. 12-29. We could now use either

$$a^2 = b^2 + c^2 - 2bc \cos A$$

or

$$\frac{a}{\sin A} = \frac{b}{\sin B}$$

to find the measure of angle A. Since the law of sines requires less work, use it to find the measure of angle A.

$$\frac{10}{\sin A} = \frac{9.9209}{\sin 49°}$$

$$\frac{10}{\sin A} = \frac{9.9209}{.75471}$$

$$9.9209 \sin A = (10)(.75471)$$

$$\sin A = \frac{7.5471}{9.9209}$$

$$= .76073$$

angle $A \approx 49°32'$

60. Refer to Fig. 12-29. Since angle A is $49°32'$ and angle B is $49°$, what is the measure of angle C?

angle $C = 180° - 49°32' - 49°$
$= 81°28'$

$$\begin{array}{r} 179°60' \\ - 98°32' \\ \hline 81°28' \end{array}$$

61. In solving a triangle that is not a right triangle,

> (a) The *law of sines* is used if you know two sides and the angle opposite one of the sides or if you know two angles and a side opposite one of these angles.
>
> (b) The *law of cosines* is used if you know the three sides or two sides and the included angle; any of the three formulas
> $$a^2 = b^2 + c^2 - 2bc \cos A$$
> $$b^2 = a^2 + c^2 - 2ac \cos B$$
> $$c^2 = a^2 + b^2 - 2ab \cos C$$
> can be used first if the three sides are known, while the formula that has the known angle in it must be used first if you know two sides and the included angle.

Solve the triangle in Fig. 12-30.

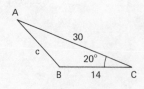

FIGURE 12-30

$c = 17.512,$
$\measuredangle A = 15°52'$
$\measuredangle B = 144°8'$
$[c^2 = a^2 + b^2 - 2ab \cos C$
$c^2 = (14)^2 + (30)^2 - 2(14)(30) \cos 20°$
$c^2 = 196 + 900 - 840(.93969)$
$c^2 = 1096 - 789.3396$
$c^2 = 306.6604$
$c \approx 17.512$
Then use the law of sines.]

Exercise Set 12-4.   Work all the problems before checking your answers.

1. Give the three versions of the law of cosines.

    _____

    _____

    _____

2. Find the measure of angle A, if
    (a) $\cos A = {}^-.90631$
    (b) $\cos A = .89101$
    (c) $\cos A = {}^-.45399$

3. Solve the triangles.
    (a)

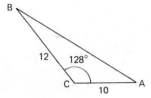

    (b) $a = 50, b = 49, c = 52$

SELF-TEST

1. Define the following functions for right angles:
    (a) sine
    (b) cosine
    (c) tangent
    (d) cotangent
    (e) secant
    (f) cosecant

2. Find $\sin A$, $\cos A$, $\sec B$ in the triangle in Fig. 12-31.

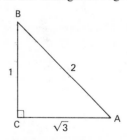

FIGURE 12-31

3. Solve the right triangle in Fig. 12-32.

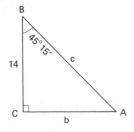

FIGURE 12-32

4. Solve the right triangle in Fig. 12-33.

_____

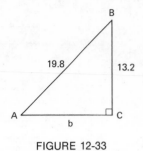

FIGURE 12-33

5. State the law of sines.

_____

6. $\sin (180° - \theta) = \sin$ _____

_____

7. $\sin 107° =$

_____

8. $\tan 93° =$

_____

9. $\cos 165° =$

_____

10. Solve the triangle in Fig. 12-34.

$\measuredangle A =$ _____

$\measuredangle B =$ _____

$b =$ _____

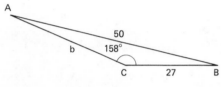

FIGURE 12-34

11. Give the three versions of the law of cosines:

_____

_____

_____

12. Find the measure of angle A if $\cos A = {}^-.89879$.

_____

13. Solve the triangle in Fig. 12-35.

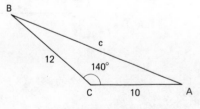

FIGURE 12-35

APPLIED PROBLEMS

A. Consider the diagram in Fig. 12-36. If the holes are to be drilled the same distance apart around the circle, $AB = BC$, etc. Also, since OA, OB, and OC are radii of the circle, they are all equal. Thus, triangles OAB and OBC are congruent and $\measuredangle OBA = \measuredangle OBC$. In the same way it can be shown that ∡*OAB* and ∡*OCB* have the same measure as ∡*OBA* and ∡*OBC*—one-half of the interior angle, ∡*ABC*. From Chapter 3, Problem D, the interior angles of an *n*-sided regular polygon equal 180° − (360°/*n*). Find the missing values in the table below.

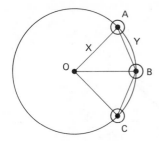

FIGURE 12-36

	n	∡*OBA*	∡*AOB*	*X*	*Y*
1.	5			1″	
2.	6			2″	
3.	8			1.5″	
4.	9			3″	
5.	10			7″	
6.	12			8.5″	

Example: 1.   The interior angles of a five-sided regular polygon are 180° − (360°/5) = 180° − 72° = 108°. ∡*OBA* = ½(108°) = 54°. Thus, ∡*OAB* = 54° so ∡*AOB* = 180° − 54° − 54° = 72°. By the law of sines, 1″/sin 54° = *Y*/sin 72°; solving for *Y*, you find *Y* ≈ 1.1756″.

B.   Use the law of cosines to find the angles of a wedge with sides of the following length.

	a	*b*	*c*	∡*A*	∡*B*	∡*C*
1.	4″	4″	3″			
2.	7″	7″	4″			
3.	5″	5″	2″			
4.	10″	10″	3″			

C.   To check the accuracy of the angle, *θ*, of the male dovetail in Fig. 12-37, you measure the distance, *X*, across the plugs of diameter *Y*. If the angle was not accurately cut, the distance, *X*, will be either too large (the angle, *θ*, is too small) or too small (the angle, θ, is too large). In order to use this technique you must be able to determine what X should be if the angle was accurately cut. Determine X for the given θ and Y.

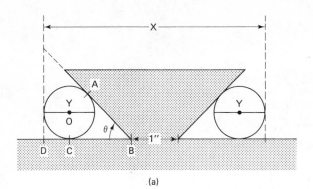

(a)

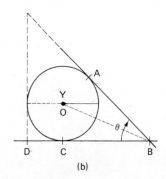

(b)

FIGURE 12-37

	θ	Y	X
1.	60°	$\frac{1}{2}''$	
2.	45°	$\frac{1}{2}''$	
3.	55°	$\frac{3}{4}''$	
4.	30°	$\frac{3}{4}''$	
5.	60°	$1''$	

Example: 1. $\angle OBC = \frac{1}{2}\theta = \frac{1}{2}(60°) = 30°$. (Why?)
$OC = \frac{1}{2}(Y) = \frac{1}{2}(\frac{1}{2}'') = \frac{1}{4}''$, since it is a radius. $\tan \angle OBC = OC/BC$, so $\tan 30° = \frac{1}{4}''/BC$ and $.57735 = \frac{1}{4}''/BC$, so $BC = .43301''$. $CD = \frac{1}{2}(\frac{1}{2}'') = \frac{1}{4}''$. (Why?) Thus, $BO = BC + CD = .43301'' + .25'' = .68301''$. This distance is on both sides, so $X = .68301'' + .68301'' + 1'' = 2.36601''$.

D. See Fig. 12-38. Use the law of cosines and law of sines to show
1. If the corresponding sides of two triangles have the same length, they are congruent (i.e., the corresponding angles are also equal). (*Hint:*

Call the triangles *ABC* and *DEF*. Given: $AB = DE$, $BC = EF$, $AC = DF$. Use the law of cosines to show $\angle ABC = \angle DEF$. Then use the law of sines.)

2. If two sides and the included angle of one triangle are the same as two sides and the included angle of a second triangle, the triangles are congruent.

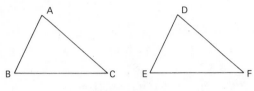

FIGURE 12-38

E. Using the definitions for the sine, tangent, and cosine of an angle in a right triangle *ABC* with sides *a*, *b*, and *c* where *a* is opposite angle *A*, *b* is opposite angle *B*, and *c* is the hypotenuse, prove the following identities for right triangles.

1. $\tan A = \dfrac{\sin A}{\cos A}$

2. $\sin^2 A + \cos^2 A = 1$

3. $\sin A = \cos B = \cos (90° - A)$

F. The area of the sector, A_S, is equal to the area of the (shaded) segment, *A*, plus the area of the triangle, A_T, in Fig. 12-39. Find the missing values in the table below. [Recall from Chapter 11, Problem B, that $A_S/A_C = \theta/360°$ and from Chapter 8, Problem G, that $A_T = \sqrt{s(s - a)(s - b)(s - c)}.$]

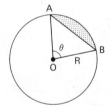

FIGURE 12-39

	θ	R	AB	A_S	A_T	A
1.	60°	3″				
2.	45°	4″				
3.	90°	7″				
4.	30°	4″				
5.	100°	6″				
6.	20°	8″				

Example: 1.   Using the law of cosines,

$$AB^2 = 3^2 + 3^2 - 2(3)(3) \cos 60°$$

so $AB^2 = 9$ and $AB = 3''$.

$$\frac{A_S}{A_C} = \frac{60°}{360°} \quad \text{and} \quad A_C = \pi(3^2) \approx 28.3 \text{ in.}^2$$

Thus,

$$A_S = \frac{28.3 \text{ in.}^2}{6} = 4.71 \text{ in.}^2$$

$$A_T = \sqrt{4.5(1.5)(1.5)(1.5)} \approx 3.90 \text{ in.}^2 \qquad A = 4.71 \text{ in.}^2 - 3.90 \text{ in.}^2 = .81 \text{ in.}^2$$

G.   Refer to Problem K in Chapter 11. By dividing the tract of land into triangles as shown in Fig. 12-40, and using the law of cosines and the formula

$$A = \sqrt{s(s - a)(s - b)(s - c)}$$

from Problem G in Chapter 8 for the area of a triangle, you can find the area of the tract of land if you know the distances *AB*, *BC*, *CD*, *DE*, and *AE*.

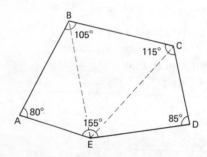

FIGURE 12-40

	AB	*BC*	*CD*	*DE*	*AE*	*Area*
1.	200′	250′	125′	180′	100′	
2.	75 m	100 m	50 m	70 m	40 m	
3.	300′	325′	170′	230′	150′	
4.	550.2′	623.4′	450′	520′	412′	
5.	68 m	95.4 m	53.2 m	65 m	48.3 m	
6.	180 m	195.3 m	161 m	172 m	135 m	

Example: 1.   First consider triangle *ABE*. By the law of cosines,

$$BE^2 = AB^2 + AE^2 - 2(AB)(AE) \cos (\angle BAE)$$
$$= (200)^2 + (100)^2 - 2(200)(100) \cos 80°$$
$$= 40,000 + 10,000 - (40,000)(.17365)$$
$$= 50,000 - 6,946$$
$$BE^2 = 43,054$$
$$BE = \sqrt{43,054} \approx 207$$

Using $A = \sqrt{s(s-a)(s-b)(s-c)}$ with $s = 507$, we find

$$A = \sqrt{(507)(307)(407)(300)}$$
$$\approx 138,000 \text{ sq ft}$$

In the same way, you can find that in triangle *CDE*, *CE* $\approx$ 210 and the area of the triangle is approximately 143,000 sq ft. Finally, we know the three sides of triangle *BCE* and the formula gives an approximate area of 242,000. Thus, the area of the tract of land is approximately 523,000 square feet.

H.   You are given the diagram in Fig. 12-41. Find the dimensions missing in the table:

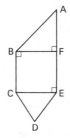

FIGURE 12-41

	AB	*BC*	*CD*	*DE*	*AE*	*AF*	*BF*	$\angle CDE$
1.	5″	6″	4″	4″	10″			
2.	10″	8″	6″		15″			60°
3.	13 cm	9 cm	6 cm	7 cm		5 cm		
4.	6″	7″	5″	4″	11″			
5.	15 cm	11 cm	7 cm	8 cm	13 cm			

I. Find the length of side *AC* in Fig. 12-42 given the data in the table:

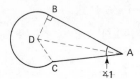

FIGURE 12-42

	AB	DC	⦟1	AC
1.	4.00″	1.00″	20°	
2.	20 cm	5 cm	25°	
3.	88 mm	30 mm	30°	
4.	2.3″	.45″	15°45′	
5.	17″	5″	32°10′	

J. The height of a building is *H* feet if the angle (at a distance of *X* feet from the building) from the ground to the top of the building is *Y* degrees. Find the missing dimensions in the table.

	H	X	Y
1.		50′	74°
2.	600′	30′	
3.	750′		60°
4.	220′	100′	
5.		60′	80°

K. Copy triangle *ABC* on a separate piece of paper and cut out the triangle. Then wrap the triangle around a pen or pencil starting with the shortest side as indicated. The result is a helix. To find the helix angle, ⦟ *CAB* in triangle *ABC*, we use the fact that the tangent of the helix angle is equal to the lead, *BC*, divided by the circumference (*AC* in the figures), so tan ⦟ *CAB* = *BC*/*AC*. Find the helix angle for the following leads and circumferences:

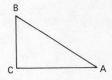

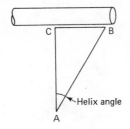

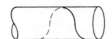

	BC	AC	∡ CAB
1.	.1″	1″	
2.	.125″	1.5″	
3.	.2″	1.3″	
4.	1 cm	5 cm	
5.	1.2 cm	4.3 cm	

answer key

PRETEST 1

1. (a) 3; (b) proper
2. (a) $\frac{80}{11}$; (b) $8\frac{2}{3}$
3. (a) $1, 2, 3, 4, 6, 12$; (b) no; (c) $2 \times 3 \times 3 \times 3 \times 3$
4. (a) $\frac{5}{7}$; (b) $\frac{30}{162}$
5. $\frac{11}{5} = 2\frac{1}{5}$
6. (a) $\frac{4}{13}$; (b) $\frac{1}{5}$
7. $\frac{20}{27}$
8. 540
9. (a) $\frac{71}{90}$; (b) $\frac{13}{28}$
10. (a) $7\frac{1}{9}$; (b) $2\frac{1}{8}$

EXERCISE SET 1-1

1. 5, 6
2. 7, 5
3. (a) Proper; (b) Proper;
   (c) Improper; (d) Improper;
4. $1\frac{7}{8}$

EXERCISE SET 1-2

1. (a) $\frac{47}{8}$; (b) $\frac{52}{15}$; (c) $\frac{30}{17}$
2. (a) $1\frac{3}{7}$; (b) 4; (c) $5\frac{4}{7}$

EXERCISE SET 1-3

1. $1, 2, 3, 4, 6, 8, 12, 24$
2. (a) Prime; (b) Composite; (c) Composite
3. (a) 5×5; (b) 17;
   (c) $5 \times 5 \times 5$ (d) $2 \times 2 \times 3 \times 7$

EXERCISE SET 1-4

1. (a) $\frac{3}{7}$; (b) $\frac{1}{6}$; (c) $\frac{7}{11}$; (d) $\frac{1}{5}$
2. (a) $\frac{136}{164}$; (b) $\frac{88}{96}$; (c) $\frac{150}{405}$; (d) $\frac{63}{168}$

EXERCISE SET 1-5

1. (a) $\frac{18}{55}$; (b) $\frac{1}{7}$; (c) $\frac{1}{14}$; (d) $16\frac{5}{7}$; (e) $29\frac{1}{2}$; (f) $5\frac{1}{3}$

EXERCISE SET 1-6

1. $\frac{9}{1}$
2. (a) $\frac{1}{10}$; (b) 4; (c) $\frac{13}{11}$ or $1\frac{2}{11}$; (d) $\frac{3}{7}$

EXERCISE SET 1-7

1. $\frac{8}{21}$
2. $\frac{5}{2} = 2\frac{1}{2}$
3. $\frac{1}{16}$
4. $\frac{27}{98}$
5. $2\frac{1}{4}$
6. $1\frac{31}{44}$
7. $\frac{3}{4}$

EXERCISE SET 1-8

1. 80
2. 36
3. 210
4. 600
5. 110

EXERCISE SET 1-9

1. $\frac{17}{80}$
2. $\frac{8}{9}$
3. $\frac{29}{245}$
4. $\frac{29}{78}$
5. $\frac{5}{9}$
6. $\frac{11}{18}$
7. $\frac{5}{91}$
8. $\frac{101}{1078}$

EXERCISE SET 1-10

1. (a) $6\frac{3}{20}$; (b) $9\frac{7}{48}$; (c) $9\frac{2}{9}$
2. (a) $\frac{13}{20}$; (b) $2\frac{6}{7}$; (c) $2\frac{1}{11}$

SELF-TEST (CHAPTER 1)

1. 7, 8
2. 16, 3
3. (a) Proper; (b) Improper; (c) Improper
4. 1, 2, 3, 4, 6, 9, 12, 18, 36
5. composite
6. (a) $3 \times 5 \times 5$; (b) $3 \times 7 \times 7$; (c) 19
7. (a) $\frac{2}{3}$; (b) $\frac{2}{5}$; (c) $\frac{1}{5}$
8. (a) $\frac{84}{156}$; (b) $\frac{45}{54}$
9. $\frac{22}{7}$
10. $5\frac{1}{2}$
11. (a) $\frac{35}{48}$; (b) $\frac{7}{27}$; (c) $7\frac{1}{3}$; (d) 6
12. (a) $\frac{9}{5}$; (b) $\frac{1}{6}$
13. (a) $\frac{5}{4}$; (b) $\frac{98}{375}$; (c) $\frac{63}{100}$; (d) 18; (e) $\frac{3}{4}$
14. (a) 90; (b) 286
15. (a) $\frac{6}{7}$; (b) $\frac{89}{225}$; (c) $5\frac{1}{4}$; (d) 11; (e) $8\frac{1}{56}$
16. (a) $\frac{4}{13}$; (b) $\frac{22}{245}$; (c) $\frac{5}{8}$; (d) $2\frac{1}{2}$; (e) $3\frac{1}{3}$; (f) $\frac{59}{60}$

PRETEST 2

1. (a) (i) tens
      (ii) hundredths
      (iii) ten-thousandths

2. (a) 72,000; (b) .12; (c) 4.1
3. (a) 9.813; (b) 24.84
4. (a) .1116; (b) 50
5. (a) 160; (b) .059; (c) 450
6. (a) 0.1875; (b) 1.31

EXERCISE SET 2-1

1. 4 ten-thousands + 5 thousands + 2 hundreds + 1 ten + 3 ones
2. (a) thousands; (b) ones; (c) tenths; (d) thousandths; (e) hundred-thousandths; (f) tens; (g) ten-thousandths

EXERCISE SET 2-2

1. (a) 2,400; (b) 40,000; (c) .03; (d) 320; (e) 6.123; (f) 3700; (g) .0468; (h) 3.200
2. (a) 2; (b) 4; (c) 1

EXERCISE SET 2-3

1. (a) 9.675; (b) 1.5; (c) 9.9845; (d) 7.2
2. (a) 3.838; (b) .487; (c) 3.582

EXERCISE SET 2-4

1. (a) 1.2; (b) .011729; (c) 200.096
2. (a) 630; (b) .00005; (c) 13.0

EXERCISE SET 2-5

1. (a) 129.03; (b) .14; (c) .87; (d) .20
2. (a) 80; (b) .0021; (c) 3.0; (d) .1490

EXERCISE SET 2-6

1. (a) 1.125; (b) .1875
2. (a) 1.231; (b) .556; (c) .143

SELF-TEST (CHAPTER 2)

1. (a) ones; (b) hundreds; (c) ten-thousandths; (d) tenths; (e) hundredths
2. (a) 1.2; (b) 8.425
3. (a) 2.64; (b) 4.09; (c) 3.022
4. (a) 6,800; (b) 71; (c) .350; (d) 40.1
5. (a) .015; (b) 20; (c) 8; (d) 3.66
6. (a) 13.0104; (b) 0.093
7. (a) 140; (b) 0.00005
8. 1.22

PRETEST 3

1. (a) 1; 25; (b) 33°14′52″
2. .327″
3. 1.286″

EXERCISE SET 3-1

1. $\frac{1}{360}$ of the angle at the center of a circle traversed in making one revolution around a circle.
2. 1; 15

3. 60
4. 120
5. 4; 5
6. (a) 80°25'35''; (b) 1°10'34''
7. (a) 18°6'19''; (b) 50°18'8''; (c) 23°54'20''

EXERCISE SET 3-2

1. .025
2. .267''

EXERCISE SET 3-3

1. .004''
2. 3.129''
3. 42°25'

SELF-TEST CHAPTER 3

1. Degree: $\frac{1}{360}$ of the angle at the center of a circle
traversed in making one revolution around a circle.
Minute: $\frac{1}{60}$ of a degree.
Second: $\frac{1}{60}$ of a minute.
2. 2; 18
3. 120
4. 600
5. 18°2'12''
6. 110°5'
7. 19°14'
8. 26°59'19''
9. .025
10. .269''
11. 3.269''
12. .001''
13. .029''
14. 31°40'

PRETEST 4

1. (a) 7; (b) 6

2.

3. (a) ⁻12; (b) ⁻7
4. (a) ⁻10; (b) 11; (c) ⁻7; (d) ⁻9
5. (a) ⁻76; (b) 27; (c) ⁻3

EXERCISE SET 4-1

1. (a) positive six; (b) negative four
2. (a) ⁻6; (b) ⁺5; (c) ⁻16
3. (a) 17; (b) 0; (c) 41

EXERCISE SET 4-2

1.

2. A: ⁻8
B: ⁻1
C: +4

EXERCISE SET 4-3

1. ⁻9
2. ⁻1
3. 4
4. ⁻47
5. ⁻30
6. 5

EXERCISE SET 4-4

1. ⁻8
2. 3
3. 10
4. ⁻18
5. ⁻7

EXERCISE SET 4-5

1. 42
2. ⁻72
3. ⁻15
4. 13
5. ⁻84
6. ⁻7

SELF-TEST (CHAPTER 4)

1. (a) 6; (b) ⁻5; (c) ⁻3
2. (a) 6; (b) 0; (c) 7
3. negative thirteen

4.

5. A: ⁺4; B: ⁻8
6. (a) ⁻16; (b) ⁻3; (c) +5
7. (a) ⁻11; (b) ⁻3; (c) ⁻6; (d) 12
8. (a) ⁻84; (b) 48; (c) ⁻24; (d) ⁻3; (e) 10; (f) ⁻13

PRETEST 5

1. (a) ⁻x; (b) $\dfrac{^-4x}{7}$

2. $\frac{7}{6}$

3. $\dfrac{^-45}{4}$

4. $x = \dfrac{2}{7} + \dfrac{k}{7}$

EXERCISE SET 5-1

1. $5x$
2. ⁻$10x$
3. ⁻$2x$
4. $3x$
5. $4x$
6. $20x$
7. ⁻$6x$
8. $20x$

9. x

10. $\dfrac{^-2}{5}x$

EXERCISE SET 5-2

1. 6
2. $^-2$
3. 1
4. 9
5. $\frac{7}{3}$
6. $\frac{5}{7}$
7. $\dfrac{^-7}{5}$
8. $\dfrac{^-8}{5}$
9. 1
10. $\frac{5}{8}$

EXERCISE SET 5-3

1. $^-9$
2. $\dfrac{^-3}{14}$
3. $\dfrac{^-5}{7}$
4. $\dfrac{^-6}{7}$
5. 18
6. $\frac{15}{14}$
7. $\dfrac{^-177}{10}$
8. $\frac{44}{9}$
9. $\frac{1}{3}$
10. 5

EXERCISE SET 5-4

1. $y = 9 - x$
2. $x = \dfrac{k}{3}$
3. $x = \dfrac{y}{k}$
4. $y = 2x - 7$
5. $r = \dfrac{d}{t}$
6. $k = xy + tx$
7. $x = \dfrac{3}{y} + \dfrac{k}{y}$
8. $y = \dfrac{xn}{m}$
9. $y = \dfrac{^-k}{3} + \dfrac{x}{3}$

10. $s = \dfrac{6c}{\pi r}$

11. $Z = \dfrac{E}{I}$

SELF-TEST (CHAPTER 5)

1. (a) ^-5x; (b) ^-8x (c) ^-6x; (d) ^-x

2. (a) $^-3$; (b) 9; (c) 4; (d) $\dfrac{^-9}{2}$; (e) $^-1$; (f) $\dfrac{^-5}{8}$; (g) $\frac{5}{3}$

    (h) $^-28$; (i) $\dfrac{^-2}{7}$; (j) $\frac{28}{3}$; (k) 10; (l) $x = 6 + y$

    (m) $x = \dfrac{m}{y}$; (n) $x = \dfrac{k}{r}$; (o) $x = \dfrac{5}{y} + \dfrac{k}{y}$

PRETEST 6

1. 9:1
2. (a) $x = \frac{1}{28}$; (b) $x = 8\frac{1}{3}$
3. $10\frac{1}{2}$ hours

EXERCISE SET 6-1

1. $\frac{9}{2}$ or 9:2
2. 1.75:1
3. 5:1

EXERCISE SET 6-2

1. $A = 21$
2. $A = 1$
3. $A = 3\frac{1}{2}$

EXERCISE SET 6-3

1. $21\frac{3}{7}$ hours
2. $49.00

SELF-TEST (CHAPTER 6)

1. 1,250/3
2. 2.5:1
3. 12:1
4. (a) $A = 4\frac{2}{3}$; (b) $A = .272$; (c) $A = 6$
5. 16.8 lb
6. 20 inches

PRETEST 7

1. (a) (i) 70%
        (ii) 204%
        (iii) .31%
      (b) $\frac{5}{8}$; (c) .036
2. $16\frac{2}{3}$
3. $3,000.00

EXERCISE SET 7-1

1. (a) 60%; (b) $71\frac{3}{7}\%$; (c) $316\frac{2}{3}\%$
2. (a) 70%; (b) 103%; (c) .4%

3. (a) .4; (b) 1.34; (c) .06; (d) .034

4. (a) $\frac{4}{25}$; (b) $\frac{9}{4}$; (c) $\frac{1}{12}$

EXERCISE SET 7-2

1. $\dfrac{r}{100} = \dfrac{p}{b}$

2. 3.528

3. 16

4. $22\frac{2}{9}\%$

EXERCISE SET 7-3

1. .0675″

2. 15%

3. 200 pounds

4. 696

5. 300

SELF-TEST (CHAPTER 7)

1. (a) $31\frac{1}{4}\%$; (b) 30%; (c) $211\frac{1}{9}\%$

2. (a) 3.5%; (b) 30%; (c) 300%

3. (a) .32; (b) .0002; (c) 2.3; (d) .04

4. (a) $\frac{7}{50}$; (b) $\frac{7}{2}$; (c) $\frac{29}{250}$

5. 2.432

6. $54\frac{6}{11}$

7. $127\frac{3}{11}\%$

8. 70%

9. 11.7 pounds

10. 1.75″

11. 2.06″

PRETEST 8

1. 64

2. x^4; (b) x^{12}

3. (a) 1; (b) $\dfrac{1}{x^3}$

4. $^-4$

EXERCISE SET 8-1

1. (a) 2^4; (b) 5^5; (c) k^3

2. (a) $5 \cdot 5 \cdot 5 \cdot 5$; (b) $7 \cdot 7 \cdot 7 \cdot 7 \cdot 7 \cdot 7 \cdot 7$; (c) $x \cdot x \cdot x \cdot x \cdot x$

3. (a) 4; (b) 625; (c) 7; (d) 216

EXERCISE SET 8-2

1. 3^9

2. 5^{11}

3. 7^6

4. x^4

5. x^5

6. 5^{10}

7. x^{15}

EXERCISE SET 8-3

1. 1

2. $\dfrac{1}{x^5}$

3. 1

4. $\dfrac{1}{7^3} = \dfrac{1}{343}$

5. $\dfrac{1}{x^5}$ or x^{-5}

6. 3^{-3} or $\dfrac{1}{3^3}$

EXERCISE SET 8-4

1. 5

2. 11

3. -9

4. (a)

5. $^-4$

6. 5

SELF-TEST (CHAPTER 8)

1. 4^6

2. x^5

3. $x \cdot x \cdot x \cdot x \cdot x \cdot x \cdot x$

4. (a) $3 \cdot 3 \cdot 3 \cdot 3 = 81$; (b) 2^{12}; (c) x^4; (d) 6^{14}

5. (a) 1; (b) $a^{-6} = \dfrac{1}{a^6}$

6. $\dfrac{1}{x^4}$

7. (a) 9; (b) $^-7$; (c) $^-3$

8. yes

PRETEST 9

1. .00095

2. .09

3. (a) 1,500,000; (b) .061

4. (a) 540,640; (b) 2.272

5. 56

EXERCISE SET 9-1

1. (a) 1,000 or 10^3; (b) $\dfrac{1}{1,000}$ or 10^{-3}; (c) $\dfrac{1}{1,000,000}$ or 10^{-6}; (d) $\dfrac{1}{100}$ or 10^{-2}

2. (a) km; (b) m; (c) cm; (d) mm

3. (a) 1,000; (b) 50; (c) .008; (d) .000009

EXERCISE SET 9-2

1. 100

2. 20,000

3. .0005
4. 9.5
5. 8,000
6. 5,000
7. .48
8. 8,500
9. .1

EXERCISE SET 9-3

1. .035
2. .9
3. 400
4. 7,500,000
5. 350,000
6. .00049

EXERCISE SET 9-4

1. 55.774
2. 15.54
3. 547
4. 1.181
5. 559
6. 23.01
7. .9144
8. 57.91
9. 23.3
10. 1.24
11. 7.41
12. .0760
13. 62
14. .4047
15. 29.3
16. 30,320
17. 251
18. 4.65
19. 122
20. 1.321
21. .00305
22. .03924
23. 3,050
24. .0095103
25. 14.19
26. 2.549
27. 4,914,900
28. .4636

EXERCISE SET 9-5

1. 3 mm × 5 cm
2. 150 mm^2 or 1.5 cm^2
3. $\frac{1}{2}$ gallon of milk

SELF-TEST (CHAPTER 9)

1. (a) 10^{-3}; (b) 10^3; (c) 10^2
2. 3,000
3. .0001

4. 21.75
5. 26.52
6. 50,000
7. .0035
8. 15,000
9. .002
10. 3.88
11. 62
12. .4047
13. 1,250
14. 90.5
15. 29,000
16. .49
17. 195
18. 31.701
19. 7.570
20. 2.549
21. .0073969
22. .0549
23. 50 liters
24. 50 cm^3
25. 6 dm^2 or 600 cm^2

PRETEST 10

1. 540 rpms
2. 750 yards

EXERCISE SET 10-1

1. 525 rpms
2. $888\frac{8}{9}$ rpms

EXERCISE SET 10-2

1. $13\frac{1}{3}$ hours

SELF-TEST (CHAPTER 10)

1. ≈ 585 rpms
2. ≈ 471 rpms
3. 120 pounds

PRETEST 11

1. (a) parallel; (b) 0
2. $A = \frac{1}{2} bh$
3. 4,189 cu in.
4. $\sqrt{41}$ inches

EXERCISE SET 11-1

1. 1
2. point
3. 2
4. 3
5. plane
6. circle
7. sphere

8. 90

9. angle

10. 180

11. acute angle—angle of less than 90°
obtuse angle—angle of greater than 90°

12. parallel

13.

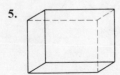

14. 3

15. A line from the center of a circle to a point on the circle.

16. Four-sided figure with exactly two sides parallel.

17. a ball, orange, etc.

18. block of wood, etc.

EXERCISE SET 11-2

1. 78

2. 9 square inches

3. 30

4. 153.94

5. 254.47

EXERCISE SET 11-3

1. 72

2. 343 cubic inches

3. 113.10 cubic inches

4. 226.20 cubic inches

5. 226.20 cubic inches

EXERCISE SET 11-4

1. $c^2 = a^2 + b^2$

2. $b = 8$

3. $c = 13$

4. $\sqrt{72} \approx 8.4853$

SELF-TEST (CHAPTER 11)

1. (a) 2; (b) 0; (c) 1; (d) 3

2. 90

3. (a) acute—angle of less than 90 degrees
(b) obtuse—angle of greater than 90 degrees

4. Two lines are perpendicular, if they form a right angle.

5.

6. (a) 49 in.^{2}; (b) ≈ 113 in.^{2}; (c) $\frac{27}{2}$ or $13\frac{1}{2}$ in^2

7. (a) 200; (b) 523.60

8. $\sqrt{300} \approx 17.3$

PRETEST 12

1. (a) $\text{sine} = \dfrac{\text{opposite side}}{\text{hypotenuse}}$

$\text{tangent} = \dfrac{\text{opposite side}}{\text{adjacent side}}$

$\text{secant} = \dfrac{\text{hypotenuse}}{\text{adjacent side}}$

(b) $\cos A = \dfrac{a}{10}$

$\cot B = \dfrac{b}{a}$

$\csc A = \dfrac{10}{b}$

(c) $\frac{3}{4}$

2. $a = 5; b \approx 8.66;$ angle $A = 60°$

3. (a) .81915
(b) $^-$.70711
(c) $B = 109°24'$
$C = 32°36'$
$b \approx 12.26$

4. (a) $\angle A = 155°$
(b) $\angle A = 59°18'$
$\angle B = 52°1'$
$\angle C = 68°41'$

EXERCISE SET 12-1

1. $\text{sine} = \dfrac{\text{opposite}}{\text{hypotenuse}}$

$\text{cosine} = \dfrac{\text{adjacent}}{\text{hypotenuse}}$

$\text{tangent} = \dfrac{\text{opposite}}{\text{adjacent}}$

$\text{secant} = \dfrac{\text{hypotenuse}}{\text{adjacent}}$

$\text{cosecant} = \dfrac{\text{hypotenuse}}{\text{opposite}}$

$\text{cotangent} = \dfrac{\text{adjacent}}{\text{opposite}}$

2. $\sin A = \dfrac{3}{3\sqrt{2}} = \dfrac{1}{\sqrt{2}}$

$\tan B = \dfrac{3}{3} = 1$

$\sec B = \dfrac{3\sqrt{2}}{3} = \sqrt{2}$

3. $\csc A = \dfrac{\sqrt{29}}{5}$

$\cos B = \dfrac{5}{\sqrt{29}}$

$\cot B = \frac{5}{2}$

4. $\frac{5}{7}$

EXERCISE SET 12-2

1. $\angle A = 60°$
   $\angle B = 30°$
   $a = 5.1962$
2. $\angle B = 55°$
   $a = 6.8830$
   $b = 9.8298$

EXERCISE SET 12-3

1. θ
2. .68200
3. $^-.5$
4. $^-7.1154$
5. $\angle C = 20°37'$
   $\angle B = 134°23'$
   $b = 50.732$

EXERCISE SET 12-4

1. $a^2 = b^2 + c^2 - 2bc \cos A$
   $b^2 = a^2 + c^2 - 2ac \cos B$
   $c^2 = a^2 + b^2 - 2ab \cos C$
2. (a) $A = 155°$
   (b) $A = 27°$
   (c) $A = 117°$
3. (a) $c = 19.793$
      $\angle B = 23°28'$
      $\angle A = 28°32'$
   (b) $\angle A = 59°16'$
      $\angle B = 57°22'$
      $\angle C = 63°22'$

SELF-TEST (CHAPTER 12)

1. (a) $\dfrac{\text{opposite side}}{\text{hypotenuse}}$

   (b) $\dfrac{\text{adjacent side}}{\text{hypotenuse}}$

   (c) $\dfrac{\text{opposite side}}{\text{adjacent side}}$

   (d) $\dfrac{\text{adjacent side}}{\text{opposite side}}$

   (e) $\dfrac{\text{hypotenuse}}{\text{adjacent side}}$

   (f) $\dfrac{\text{hypotenuse}}{\text{opposite side}}$

2. $\sin A = \frac{1}{2}$
   $\cos A = \dfrac{\sqrt{3}}{2}$
   $\sec B = \dfrac{2}{\sqrt{3}}$
3. $\angle A = 44°45'$
   $b = 14.123 \approx 14$
   $c = 19.886 \approx 20$
4. $\angle A = 41°49'$
   $\angle B = 48°11'$
   $b \approx 14.8$
5. $\dfrac{a}{\sin A} = \dfrac{b}{\sin B} = \dfrac{c}{\sin C}$
6. θ
7. .95630
8. $^-19.081$
9. $^-.96593$
10. $\angle A = 11°40'$
    $\angle B = 10°20'$
    $b = 23.98 \approx 24$
11. $a^2 = b^2 + c^2 - 2bc \cos A$
    $b^2 = a^2 + c^2 - 2ac \cos B$
    $c^2 = a^2 + b^2 - 2ab \cos C$
12. $\angle A = 154°$
13. $\angle A = 21°54'$
    $\angle B = 18°6'$
    $c = 20.685$

METRIC CONVERSION TABLE

Metric	To	English	Conversion Factor	English	To	Metric	Conversion Factor
Meters	to	Inches	39.37 in./m	Inches	to	Millimeters	25.4 mm/in.
Meters	to	Feet	3.2808 ft/m	Feet	to	Meters	.3048 m/ft
Meters	to	Yards	1.094 yd/m	Yards	to	Meters	.9144 m/yd
Kilometers	to	Miles	.6214 mi/km	Miles	to	Kilometers	1.609 km/mi
Square centimeters	to	Square inches	$.155 \text{ in.}^2/\text{cm}^2$	Square inches	to	Square centimeters	$6.452 \text{ cm}^2/\text{in.}^2$
Square meters	to	Square feet	$10.764 \text{ ft}^2/\text{m}^2$	Square feet	to	Square meters	$.0929 \text{ m}^2/\text{ft}^2$
Square meters	to	Square yards	$1.196 \text{ yd}^2/\text{m}^2$	Square yards	to	Square meters	$.836 \text{ m}^2/\text{yd}^2$
Ares	to	Acres	.0247 acres/a	Acres	to	Ares	40.47 a/acre
Cubic centimeters	to	Cubic inches	$.0610 \text{ in.}^3/\text{cm}^3$	Cubic inches	to	Cubic centimeters	$16.383 \text{ cm}^3/\text{in.}^3$
Cubic meters	to	Cubic feet	$35.314 \text{ ft}^3/\text{m}^3$	Cubic feet	to	Cubic meters	$.02832 \text{ m}^3/\text{ft}^3$
Cubic meters	to	Cubic yards	$1.308 \text{ yd}^3/\text{m}^3$	Cubic yards	to	Cubic meters	$.7645 \text{ m}^3/\text{yd}^3$
Liters	to	U.S. quarts	1.0567 qt/ℓ	U.S. quarts	to	Liters	.9462 ℓ/qt
Liters	to	U.S. gallons	.2642 gal/ℓ	U.S. gallons	to	Liters	3.785 ℓ/gal
Liters	to	Pounds of water at 62°F	2.202 lb/ℓ				

table of
natural trigonometric functions

Values of the sine, cosine, tangent, and cotangent[1] are given for each minute of angle from 0° to 180°.

Use the column headings at the top of the page for degrees indicated at the top of the page and the column headings at the bottom for degrees indicated at the bottom.

Use the minute column at the left of each block when degrees are stated at the left (either at the top or bottom of the page). Use the minute column at the right for degrees stated on the right (either top or bottom).

The correct sign (plus or minus) must be supplied as shown in the text.

---

[1] Ctn may be used interchangeably with cot as an abbreviation for cotangent

Table of natural trigonometric functions taken from Albert L. Hoag, Donald C. McNeese, ENGINEERING HANDBOOK, © 1957, pp. 276-299. Reprinted by permission of Prentice-Hall, Inc.

0°					179°	1°					178°
′	Sin	Tan	Ctn	Cos	′	′	Sin	Tan	Ctn	Cos	′
0	.00000	.00000	∞	1.0000	60	0	.01745	.01746	57.290	.99985	60
1	.00029	.00029	3437.7	1.0000	59	1	.01774	.01775	56.351	.99984	59
2	.00058	.00058	1718.9	1.0000	58	2	.01803	.01804	55.442	.99984	58
3	.00087	.00087	1145.9	1.0000	57	3	.01832	.01833	54.561	.99983	57
4	.00116	.00116	859.44	1.0000	56	4	.01862	.01862	53.709	.99983	56
5	.00145	.00145	687.55	1.0000	55	5	.01891	.01891	52.882	.99982	55
6	.00175	.00175	572.96	1.0000	54	6	.01920	.01920	52.081	.99982	54
7	.00204	.00204	491.11	1.0000	53	7	.01949	.01949	51.303	.99981	53
8	.00233	.00233	429.72	1.0000	52	8	.01978	.01978	50.549	.99980	52
9	.00262	.00262	381.97	1.0000	51	9	.02007	.02007	49.816	.99980	51
10	.00291	.00291	343.77	1.0000	50	10	.02036	.02036	49.104	.99979	50
11	.00320	.00320	312.52	.99999	49	11	.02065	.02066	48.412	.99979	49
12	.00349	.00349	286.48	.99999	48	12	.02094	.02095	47.740	.99978	48
13	.00378	.00378	264.44	.99999	47	13	.02123	.02124	47.085	.99977	47
14	.00407	.00407	245.55	.99999	46	14	.02152	.02153	46.449	.99977	46
15	.00436	.00436	229.18	.99999	45	15	.02181	.02182	45.829	.99976	45
16	.00465	.00465	214.86	.99999	44	16	.02211	.02211	45.226	.99976	44
17	.00495	.00495	202.22	.99999	43	17	.02240	.02240	44.639	.99975	43
18	.00524	.00524	190.98	.99999	42	18	.02269	.02269	44.066	.99974	42
19	.00553	.00553	180.93	.99998	41	19	.02298	.02298	43.508	.99974	41
20	.00582	.00582	171.89	.99998	40	20	.02327	.02328	42.964	.99973	40
21	.00611	.00611	163.70	.99998	39	21	.02356	.02357	42.433	.99972	39
22	.00640	.00640	156.26	.99998	38	22	.02385	.02386	41.916	.99972	38
23	.00669	.00669	149.47	.99998	37	23	.02414	.02415	41.411	.99971	37
24	.00698	.00698	143.24	.99998	36	24	.02443	.02444	40.917	.99970	36
25	.00727	.00727	137.51	.99997	35	25	.02472	.02473	40.436	.99969	35
26	.00756	.00756	132.22	.99997	34	26	.02501	.02502	39.965	.99969	34
27	.00785	.00785	127.32	.99997	33	27	.02530	.02531	39.506	.99968	33
28	.00814	.00815	122.77	.99997	32	28	.02560	.02560	39.057	.99967	32
29	.00844	.00844	118.54	.99996	31	29	.02589	.02589	38.618	.99966	31
30	.00873	.00873	114.59	.99996	30	30	.02618	.02619	38.188	.99966	30
31	.00902	.00902	110.89	.99996	29	31	.02647	.02648	37.769	.99965	29
32	.00931	.00931	107.43	.99996	28	32	.02676	.02677	37.358	.99964	28
33	.00960	.00960	104.17	.99995	27	33	.02705	.02706	36.956	.99963	27
34	.00989	.00989	101.11	.99995	26	34	.02734	.02735	36.563	.99963	26
35	.01018	.01018	98.218	.99995	25	35	.02763	.02764	36.178	.99962	25
36	.01047	.01047	95.489	.99995	24	36	.02792	.02793	35.801	.99961	24
37	.01076	.01076	92.908	.99994	23	37	.02821	.02822	35.431	.99960	23
38	.01105	.01105	90.463	.99994	22	38	.02850	.02851	35.070	.99959	22
39	.01134	.01135	88.144	.99994	21	39	.02879	.02881	34.715	.99959	21
40	.01164	.01164	85.940	.99993	20	40	.02908	.02910	34.368	.99958	20
41	.01193	.01193	83.844	.99993	19	41	.02938	.02939	34.027	.99957	19
42	.01222	.01222	81.847	.99993	18	42	.02967	.02968	33.694	.99956	18
43	.01251	.01251	79.943	.99992	17	43	.02996	.02997	33.366	.99955	17
44	.01280	.01280	78.126	.99992	16	44	.03025	.03026	33.045	.99954	16
45	.01309	.01309	76.390	.99991	15	45	.03054	.03055	32.730	.99953	15
46	.01338	.01338	74.729	.99991	14	46	.03083	.03084	32.421	.99952	14
47	.01367	.01367	73.139	.99991	13	47	.03112	.03114	32.118	.99952	13
48	.01396	.01396	71.615	.99990	12	48	.03141	.03143	31.821	.99951	12
49	.01425	.01425	70.153	.99990	11	49	.03170	.03172	31.528	.99950	11
50	.01454	.01455	68.750	.99989	10	50	.03199	.03201	31.242	.99949	10
51	.01483	.01484	67.402	.99989	9	51	.03228	.03230	30.960	.99948	9
52	.01513	.01513	66.105	.99989	8	52	.03257	.03259	30.683	.99947	8
53	.01542	.01542	64.858	.99988	7	53	.03286	.03288	30.412	.99946	7
54	.01571	.01571	63.657	.99988	6	54	.03316	.03317	30.145	.99945	6
55	.01600	.01600	62.499	.99987	5	55	.03345	.03346	29.882	.99944	5
56	.01629	.01629	61.383	.99987	4	56	.03374	.03376	29.624	.99943	4
57	.01658	.01658	60.306	.99986	3	57	.03403	.03405	29.371	.99942	3
58	.01687	.01687	59.266	.99986	2	58	.03432	.03434	29.122	.99941	2
59	.01716	.01716	58.261	.99985	1	59	.03461	.03463	28.877	.99940	1
60	.01745	.01746	57.290	.99985	0	60	.03490	.03492	28.636	.99939	0
′	Cos	Ctn	Tan	Sin	′	′	Cos	Ctn	Tan	Sin	′
90°					89°	91°					88°

2°					177°	3°					176°
′	Sin	Tan	Ctn	Cos	′	′	Sin	Tan	Ctn	Cos	′
0	.03490	.03492	28.636	.99939	60	0	.05234	.05241	19.081	.99863	60
1	.03519	.03521	28.399	.99938	59	1	.05263	.05270	18.976	.99861	59
2	.03548	.03550	28.166	.99937	58	2	.05292	.05299	18.871	.99860	58
3	.03577	.03579	27.937	.99936	57	3	.05321	.05328	18.768	.99858	57
4	.03606	.03609	27.712	.99935	56	4	.05350	.05357	18.666	.99857	56
5	.03635	.03638	27.490	.99934	55	5	.05379	.05387	18.564	.99855	55
6	.03664	.03667	27.271	.99933	54	6	.05408	.05416	18.464	.99854	54
7	.03693	.03696	27.057	.99932	53	7	.05437	.05445	18.366	.99852	53
8	.03723	.03725	26.845	.99931	52	8	.05466	.05474	18.268	.99851	52
9	.03752	.03754	26.637	.99930	51	9	.05495	.05503	18.171	.99849	51
10	.03781	.03783	26.432	.99929	50	10	.05524	.05533	18.075	.99847	50
11	.03810	.03812	26.230	.99927	49	11	.05553	.05562	17.980	.99846	49
12	.03839	.03842	26.031	.99926	48	12	.05582	.05591	17.886	.99844	48
13	.03868	.03871	25.835	.99925	47	13	.05611	.05620	17.793	.99842	47
14	.03897	.03900	25.642	.99924	46	14	.05640	.05649	17.702	.99841	46
15	.03926	.03929	25.452	.99923	45	15	.05669	.05678	17.611	.99839	45
16	.03955	.03958	25.264	.99922	44	16	.05698	.05708	17.521	.99838	44
17	.03984	.03987	25.080	.99921	43	17	.05727	.05737	17.431	.99836	43
18	.04013	.04016	24.898	.99919	42	18	.05756	.05766	17.343	.99834	42
19	.04042	.04046	24.719	.99918	41	19	.05785	.05795	17.256	.99833	41
20	.04071	.04075	24.542	.99917	40	20	.05814	.05824	17.169	.99831	40
21	.04100	.04104	24.368	.99916	39	21	.05844	.05854	17.084	.99829	39
22	.04129	.04133	24.196	.99915	38	22	.05873	.05883	16.999	.99827	38
23	.04159	.04162	24.026	.99913	37	23	.05902	.05912	16.915	.99826	37
24	.04188	.04191	23.859	.99912	36	24	.05931	.05941	16.832	.99824	36
25	.04217	.04220	23.695	.99911	35	25	.05960	.05970	16.750	.99822	35
26	.04246	.04250	23.532	.99910	34	26	.05989	.05999	16.668	.99821	34
27	.04275	.04279	23.372	.99909	33	27	.06018	.06029	16.587	.99819	33
28	.04304	.04308	23.214	.99907	32	28	.06047	.06058	16.507	.99817	32
29	.04333	.04337	23.058	.99906	31	29	.06076	.06087	16.428	.99815	31
30	.04362	.04366	22.904	.99905	30	30	.06105	.06116	16.350	.99813	30
31	.04391	.04395	22.752	.99904	29	31	.06134	.06145	16.272	.99812	29
32	.04420	.04424	22.602	.99902	28	32	.06163	.06175	16.195	.99810	28
33	.04449	.04454	22.454	.99901	27	33	.06192	.06204	16.119	.99808	27
34	.04478	.04483	22.308	.99900	26	34	.06221	.06233	16.043	.99806	26
35	.04507	.04512	22.164	.99898	25	35	.06250	.06262	15.969	.99804	25
36	.04536	.04541	22.022	.99897	24	36	.06279	.06291	15.895	.99803	24
37	.04565	.04570	21.881	.99896	23	37	.06308	.06321	15.821	.99801	23
38	.04594	.04599	21.743	.99894	22	38	.06337	.06350	15.748	.99799	22
39	.04623	.04628	21.606	.99893	21	39	.06366	.06379	15.676	.99797	21
40	.04653	.04658	21.470	.99892	20	40	.06395	.06408	15.605	.99795	20
41	.04682	.04687	21.337	.99890	19	41	.06424	.06438	15.534	.99793	19
42	.04711	.04716	21.205	.99889	18	42	.06453	.06467	15.464	.99792	18
43	.04740	.04745	21.075	.99888	17	43	.06482	.06496	15.394	.99790	17
44	.04769	.04774	20.946	.99886	16	44	.06511	.06525	15.325	.99788	16
45	.04798	.04803	20.819	.99885	15	45	.06540	.06554	15.257	.99786	15
46	.04827	.04833	20.693	.99883	14	46	.06569	.06584	15.189	.99784	14
47	.04856	.04862	20.569	.99882	13	47	.06598	.06613	15.122	.99782	13
48	.04885	.04891	20.446	.99881	12	48	.06627	.06642	15.056	.99780	12
49	.04914	.04920	20.325	.99879	11	49	.06656	.06671	14.990	.99778	11
50	.04943	.04949	20.206	.99878	10	50	.06685	.06700	14.924	.99776	10
51	.04972	.04978	20.087	.99876	9	51	.06714	.06730	14.860	.99774	9
52	.05001	.05007	19.970	.99875	8	52	.06743	.06759	14.795	.99772	8
53	.05030	.05037	19.855	.99873	7	53	.06773	.06788	14.732	.99770	7
54	.05059	.05066	19.740	.99872	6	54	.06802	.06817	14.669	.99768	6
55	.05088	.05095	19.627	.99870	5	55	.06831	.06847	14.606	.99766	5
56	.05117	.05124	19.516	.99869	4	56	.06860	.06876	14.544	.99764	4
57	.05146	.05153	19.405	.99867	3	57	.06889	.06905	14.482	.99762	3
58	.05175	.05182	19.296	.99866	2	58	.06918	.06934	14.421	.99760	2
59	.05205	.05212	19.188	.99864	1	59	.06947	.06963	14.361	.99758	1
60	.05234	.05241	19.081	.99863	0	60	.06976	.06993	14.301	.99756	0
′	Cos	Ctn	Tan	Sin	′	′	Cos	Ctn	Tan	Sin	′

| 92° | | | | | 87° | 93° | | | | | 86° |

4°					175°		5°					174°
′	Sin	Tan	Ctn	Cos	′		′	Sin	Tan	Ctn	Cos	′
0	.06976	.06993	14.301	.99756	60		0	.08716	.08749	11.430	.99619	60
1	.07005	.07022	14.241	.99754	59		1	.08745	.08778	11.392	.99617	59
2	.07034	.07051	14.182	.99752	58		2	.08774	.08807	11.354	.99614	58
3	.07063	.07080	14.124	.99750	57		3	.08803	.08837	11.316	.99612	57
4	.07092	.07110	14.065	.99748	56		4	.08831	.08866	11.279	.99609	56
5	.07121	.07139	14.008	.99746	55		5	.08860	.08895	11.242	.99607	55
6	.07150	.07168	13.951	.99744	54		6	.08889	.08925	11.205	.99604	54
7	.07179	.07197	13.894	.99742	53		7	.08918	.08954	11.168	.99602	53
8	.07208	.07227	13.838	.99740	52		8	.08947	.08983	11.132	.99599	52
9	.07237	.07256	13.782	.99738	51		9	.08976	.09013	11.095	.99596	51
10	.07266	.07285	13.727	.99736	50		10	.09005	.09042	11.059	.99594	50
11	.07295	.07314	13.672	.99734	49		11	.09034	.00971	11.024	.99591	49
12	.07324	.07344	13.617	.99731	48		12	.09063	.09101	10.988	.99588	48
13	.07353	.07373	13.563	.99729	47		13	.09092	.09130	10.953	.99586	47
14	.07382	.07402	13.510	.99727	46		14	.09121	.09159	10.918	.99583	46
15	.07411	.07431	13.457	.99725	45		15	.09150	.09189	10.883	.99580	45
16	.07440	.07461	13.404	.99723	44		16	.09179	.09218	10.848	.99578	44
17	.07469	.07490	13.352	.99721	43		17	.09208	.09247	10.814	.99575	43
18	.07498	.07519	13.300	.99719	42		18	.09237	.09277	10.780	.99572	42
19	.07527	.07548	13.248	.99716	41		19	.09266	.09306	10.746	.99570	41
20	.07556	.07578	13.197	.99714	40		20	.09295	.09335	10.712	.99567	40
21	.07585	.07607	13.146	.99712	39		21	.09324	.09365	10.678	.99564	39
22	.07614	.07636	13.096	.99710	38		22	.09353	.09394	10.645	.99562	38
23	.07643	.07665	13.046	.99708	37		23	.09382	.09423	10.612	.99559	37
24	.07672	.07695	12.996	.99705	36		24	.09411	.09453	10.579	.99556	36
25	.07701	.07724	12.947	.99703	35		25	.09440	.09482	10.546	.99553	35
26	.07730	.07753	12.898	.99701	34		26	.09469	.09511	10.514	.99551	34
27	.07759	.07782	12.850	.99699	33		27	.09498	.09541	10.481	.99548	33
28	.07788	.07812	12.801	.99696	32		28	.09527	.09570	10.449	.99545	32
29	.07817	.07841	12.754	.99694	31		29	.09556	.09600	10.417	.99542	31
30	.07846	.07870	12.706	.99692	30		30	.09585	.09629	10.385	.99540	30
31	.07875	.07899	12.659	.99689	29		31	.09614	.09658	10.354	.99537	29
32	.07904	.07929	12.612	.99687	28		32	.09642	.09688	10.322	.99534	28
33	.07933	.07958	12.566	.99685	27		33	.09671	.09717	10.291	.99531	27
34	.07962	.07987	12.520	.99683	26		34	.09700	.09746	10.260	.99528	26
35	.07991	.08017	12.474	.99680	25		35	.09729	.09776	10.229	.99526	25
36	.08020	.08046	12.429	.99678	24		36	.09758	.09805	10.199	.99523	24
37	.08049	.08075	12.384	.99676	23		37	.09787	.09834	10.168	.99520	23
38	.08078	.08104	12.339	.99673	22		38	.09816	.09864	10.138	.99517	22
39	.08107	.08134	12.295	.99671	21		39	.09845	.09893	10.108	.99514	21
40	.08136	.08163	12.251	.99668	20		40	.09874	.09923	10.078	.99511	20
41	.08165	.08192	12.207	.99666	19		41	.09903	.09952	10.048	.99508	19
42	.08194	.08221	12.163	.99664	18		42	.09932	.09981	10.019	.99506	18
43	.08223	.08251	12.120	.99661	17		43	.09961	.10011	9.9893	.99503	17
44	.08252	.08280	12.077	.99659	16		44	.09990	.10040	9.9601	.99500	16
45	.08281	.08309	12.035	.99657	15		45	.10019	.10069	9.9310	.99497	15
46	.08310	.08339	11.992	.99654	14		46	.10048	.10099	9.9021	.99494	14
47	.08339	.08368	11.950	.99652	13		47	.10077	.10128	9.8734	.99491	13
48	.08368	.08397	11.909	.99649	12		48	.10106	.10158	9.8448	.99488	12
49	.08397	.08427	11.867	.99647	11		49	.10135	.10187	9.8164	.99485	11
50	.08426	.08456	11.826	.99644	10		50	.10164	.10216	9.7882	.99482	10
51	.08455	.08485	11.785	.99642	9		51	.10192	.10246	9.7601	.99479	9
52	.08484	.08514	11.745	.99639	8		52	.10221	.10275	9.7322	.99476	8
53	.08513	.08544	11.705	.99637	7		53	.10250	.10305	9.7044	.99473	7
54	.08542	.08573	11.664	.99635	6		54	.10279	.10334	9.6768	.99470	6
55	.08571	.08602	11.625	.99632	5		55	.10308	.10363	9.6493	.99467	5
56	.08600	.08632	11.585	.99630	4		56	.10337	.10393	9.6220	.99464	4
57	.08629	.08661	11.546	.99627	3		57	.10366	.10422	9.5949	.99461	3
58	.08658	.08690	11.507	.99625	2		58	.10395	.10452	9.5679	.99458	2
59	.08687	.08720	11.468	.99622	1		59	.10424	.10481	9.5411	.99455	1
60	.08716	.08749	11.430	.99619	0		60	.10453	.10510	9.5144	.99452	0
′	Cos	Ctn	Tan	Sin	′		′	Cos	Ctn	Tan	Sin	′

6°					173°	7°					172°
′	Sin	Tan	Ctn	Cos	′	′	Sin	Tan	Ctn	Cos	′
0	.10453	.10510	9.5144	.99452	60	0	.12187	.12278	8.1443	.99255	60
1	.10482	.10540	9.4878	.99449	59	1	.12216	.12308	8.1248	.99251	59
2	.10511	.10569	9.4614	.99446	58	2	.12245	.12338	8.1054	.99248	58
3	.10540	.10599	9.4352	.99443	57	3	.12274	.12367	8.0860	.99244	57
4	.10569	.10628	9.4090	.99440	56	4	.12302	.12397	8.0667	.99240	56
5	.10597	.10657	9.3831	.99437	55	5	.12331	.12426	8.0476	.99237	55
6	.10626	.10687	9.3572	.99434	54	6	.12360	.12456	8.0285	.99233	54
7	.10655	.10716	9.3315	.99431	53	7	.12389	.12485	8.0095	.99230	53
8	.10684	.10746	9.3060	.99428	52	8	.12418	.12515	7.9906	.99226	52
9	.10713	.10775	9.2806	.99424	51	9	.12447	.12544	7.9718	.99222	51
10	.10742	.10805	9.2553	.99421	50	10	.12476	.12574	7.9530	.99219	50
11	.10771	.10834	9.2302	.99418	49	11	.12504	.12603	7.9344	.99215	49
12	.10800	.10863	9.2052	.99415	48	12	.12533	.12633	7.9158	.99211	48
13	.10829	.10893	9.1803	.99412	47	13	.12562	.12662	7.8973	.99208	47
14	.10858	.10922	9.1555	.99409	46	14	.12591	.12692	7.8789	.99204	46
15	.10887	.10952	9.1309	.99406	45	15	.12620	.12722	7.8606	.99200	45
16	.10916	.10981	9.1065	.99402	44	16	.12649	.12751	7.8424	.99197	44
17	.10945	.11011	9.0821	.99399	43	17	.12678	.12781	7.8243	.99193	43
18	.10973	.11040	9.0579	.99396	42	18	.12706	.12810	7.8062	.99189	42
19	.11002	.11070	9.0338	.99393	41	19	.12735	.12840	7.7882	.99186	41
20	.11031	.11099	9.0098	.99390	40	20	.12764	.12869	7.7704	.99182	40
21	.11060	.11128	8.9860	.99386	39	21	.12793	.12899	7.7525	.99178	39
22	.11089	.11158	8.9623	.99383	38	22	.12822	.12929	7.7348	.99175	38
23	.11118	.11187	8.9387	.99380	37	23	.12851	.12958	7.7171	.99171	37
24	.11147	.11217	8.9152	.99377	36	24	.12880	.12988	7.6996	.99167	36
25	.11176	.11246	8.8919	.99374	35	25	.12908	.13017	7.6821	.99163	35
26	.11205	.11276	8.8686	.99370	34	26	.12937	.13047	7.6647	.99160	34
27	.11234	.11305	8.8455	.99367	33	27	.12966	.13076	7.6473	.99156	33
28	.11263	.11335	8.8225	.99364	32	28	.12995	.13106	7.6301	.99152	32
29	.11291	.11364	8.7996	.99360	31	29	.13024	.13136	7.6129	.99148	31
30	.11320	.11394	8.7769	.99357	30	30	.13053	.13165	7.5958	.99144	30
31	.11349	.11423	8.7542	.99354	29	31	.13081	.13195	7.5787	.99141	29
32	.11378	.11452	8.7317	.99351	28	32	.13110	.13224	7.5618	.99137	28
33	.11407	.11482	8.7093	.99347	27	33	.13139	.13254	7.5449	.99133	27
34	.11436	.11511	8.6870	.99344	26	34	.13168	.13284	7.5281	.99129	26
35	.11465	.11541	8.6648	.99341	25	35	.13197	.13313	7.5113	.99125	25
36	.11494	.11570	8.6427	.99337	24	36	.13226	.13343	7.4947	.99122	24
37	.11523	.11600	8.6208	.99334	23	37	.13254	.13372	7.4781	.99118	23
38	.11552	.11629	8.5989	.99331	22	38	.13283	.13402	7.4615	.99114	22
39	.11580	.11659	8.5772	.99327	21	39	.13312	.13432	7.4451	.99110	21
40	.11609	.11688	8.5555	.99324	20	40	.13341	.13461	7.4287	.99106	20
41	.11638	.11718	8.5340	.99320	19	41	.13370	.13491	7.4124	.99102	19
42	.11667	.11747	8.5126	.99317	18	42	.13399	.13521	7.3962	.99098	18
43	.11696	.11777	8.4913	.99314	17	43	.13427	.13550	7.3800	.99094	17
44	.11725	.11806	8.4701	.99310	16	44	.13456	.13580	7.3639	.99091	16
45	.11754	.11836	8.4490	.99307	15	45	.13485	.13609	7.3479	.99087	15
46	.11783	.11865	8.4280	.99303	14	46	.13514	.13639	7.3319	.99083	14
47	.11812	.11895	8.4071	.99300	13	47	.13543	.13669	7.3160	.99079	13
48	.11840	.11924	8.3863	.99297	12	48	.13572	.13698	7.3002	.99075	12
49	.11869	.11954	8.3656	.99293	11	49	.13600	.13728	7.2844	.99071	11
50	.11898	.11983	8.3450	.99290	10	50	.13629	.13758	7.2687	.99067	10
51	.11927	.12013	8.3245	.99286	9	51	.13658	.13787	7.2531	.99063	9
52	.11956	.12042	8.3041	.99283	8	52	.13687	.13817	7.2375	.99059	8
53	.11985	.12072	8.2838	.99279	7	53	.13716	.13846	7.2220	.99055	7
54	.12014	.12101	8.2636	.99276	6	54	.13744	.13876	7.2066	.99051	6
55	.12043	.12131	8.2434	.99272	5	55	.13773	.13906	7.1912	.99047	5
56	.12071	.12160	8.2234	.99269	4	56	.13802	.13935	7.1759	.99043	4
57	.12100	.12190	8.2035	.99265	3	57	.13831	.13965	7.1607	.99039	3
58	.12129	.12219	8.1837	.99262	2	58	.13860	.13995	7.1455	.99035	2
59	.12158	.12249	8.1640	.99258	1	59	.13889	.14024	7.1304	.99031	1
60	.12187	.12278	8.1443	.99255	0	60	.13917	.14054	7.1154	.99027	0
′	Cos	Ctn	Tan	Sin	′	′	Cos	Ctn	Tan	Sin	′

| 96° | | | | | 83° | 97° | | | | | 82° |

TABLE OF NATURAL TRIGONOMETRIC FUNCTIONS (Cont.)

′	Sin	Tan	Ctn	Cos	′	′	Sin	Tan	Ctn	Cos	′
0	.13917	.14054	7.1154	.99027	60	0	.15643	.15838	6.3138	.98769	60
1	.13946	.14084	7.1004	.99023	59	1	.15672	.15868	6.3019	.98764	59
2	.13975	.14113	7.0855	.99019	58	2	.15701	.15898	6.2901	.98760	58
3	.14004	.14143	7.0706	.99015	57	3	.15730	.15928	6.2783	.98755	57
4	.14033	.14173	7.0558	.99011	56	4	.15758	.15958	6.2666	.98751	56
5	.14061	.14202	7.0410	.99006	55	5	.15787	.15988	6.2549	.98746	55
6	.14090	.14232	7.0264	.99002	54	6	.15816	.16017	6.2432	.98741	54
7	.14119	.14262	7.0117	.98998	53	7	.15845	.16047	6.2316	.98737	53
8	.14148	.14291	6.9972	.98994	52	8	.15873	.16077	6.2200	.98732	52
9	.14177	.14321	6.9827	.98990	51	9	.15902	.16107	6.2085	.98728	51
10	.14205	.14351	6.9682	.98986	50	10	.15931	.16137	6.1970	.98723	50
11	.14234	.14381	6.9538	.98982	49	11	.15959	.16167	6.1856	.98718	49
12	.14263	.14410	6.9395	.98978	48	12	.15988	.16196	6.1742	.98714	48
13	.14292	.14440	6.9252	.98973	47	13	.16017	.16226	6.1628	.98709	47
14	.14320	.14470	6.9110	.98969	46	14	.16046	.16256	6.1515	.98704	46
15	.14349	.14499	6.8969	.98965	45	15	.16074	.16286	6.1402	.98700	45
16	.14378	.14529	6.8828	.98961	44	16	.16103	.16316	6.1290	.98695	44
17	.14407	.14559	6.8687	.98957	43	17	.16132	.16346	6.1178	.98690	43
18	.14436	.14588	6.8548	.98953	42	18	.16160	.16376	6.1066	.98686	42
19	.14464	.14618	6.8408	.98948	41	19	.16189	.16405	6.0955	.98681	41
20	.14493	.14648	6.8269	.98944	40	20	.16218	.16435	6.0844	.98676	40
21	.14522	.14678	6.8131	.98940	39	21	.16246	.16465	6.0734	.98671	39
22	.14551	.14707	6.7994	.98936	38	22	.16275	.16495	6.0624	.98667	38
23	.14580	.14737	6.7856	.98931	37	23	.16304	.16525	6.0514	.98662	37
24	.14608	.14767	6.7720	.98927	36	24	.16333	.16555	6.0405	.98657	36
25	.14637	.14796	6.7584	.98923	35	25	.16361	.16585	6.0296	.98652	35
26	.14666	.14826	6.7448	.98919	34	26	.16390	.16615	6.0188	.98648	34
27	.14695	.14856	6.7313	.98914	33	27	.16419	.16645	6.0080	.98643	33
28	.14723	.14886	6.7179	.98910	32	28	.16447	.16674	5.9972	.98638	32
29	.14752	.14915	6.7045	.98906	31	29	.16476	.16704	5.9865	.98633	31
30	.14781	.14945	6.6912	.98902	30	30	.16505	.16734	5.9758	.98629	30
31	.14810	.14975	6.6779	.98897	29	31	.16533	.16764	5.9651	.98624	29
32	.14838	.15005	6.6646	.98893	28	32	.16562	.16794	5.9545	.98619	28
33	.14867	.15034	6.6514	.98889	27	33	.16591	.16824	5.9439	.98614	27
34	.14896	.15064	6.6383	.98884	26	34	.16620	.16854	5.9333	.98609	26
35	.14925	.15094	6.6252	.98880	25	35	.16648	.16884	5.9228	.98604	25
36	.14954	.15124	6.6122	.98876	24	36	.16677	.16914	5.9124	.98600	24
37	.14982	.15153	6.5992	.98871	23	37	.16706	.16944	5.9019	.98595	23
38	.15011	.15183	6.5863	.98867	22	38	.16734	.16974	5.8915	.98590	22
39	.15040	.15213	6.5734	.98863	21	39	.16763	.17004	5.8811	.98585	21
40	.15069	.15243	6.5606	.98858	20	40	.16792	.17033	5.8708	.98580	20
41	.15097	.15272	6.5478	.98854	19	41	.16820	.17063	5.8605	.98575	19
42	.15126	.15302	6.5350	.98849	18	42	.16849	.17093	5.8502	.98570	18
43	.15155	.15332	6.5223	.98845	17	43	.16878	.17123	5.8400	.98565	17
44	.15184	.15362	6.5097	.98841	16	44	.16906	.17153	5.8298	.98561	16
45	.15212	.15391	6.4971	.98836	15	45	.16935	.17183	5.8197	.98556	15
46	.15241	.15421	6.4846	.98832	14	46	.16964	.17213	5.8095	.98551	14
47	.15270	.15451	6.4721	.98827	13	47	.16992	.17243	5.7994	.98546	13
48	.15299	.15481	6.4596	.98823	12	48	.17021	.17273	5.7894	.98541	12
49	.15327	.15511	6.4472	.98818	11	49	.17050	.17303	5.7794	.98536	11
50	.15356	.15540	6.4348	.98814	10	50	.17078	.17333	5.7694	.98531	10
51	.15385	.15570	6.4225	.98809	9	51	.17107	.17363	5.7594	.98526	9
52	.15414	.15600	6.4103	.98805	8	52	.17136	.17393	5.7495	.98521	8
53	.15442	.15630	6.3980	.98800	7	53	.17164	.17423	5.7396	.98516	7
54	.15471	.15660	6.3859	.98796	6	54	.17193	.17453	5.7297	.98511	6
55	.15500	.15689	6.3737	.98791	5	55	.17222	.17483	5.7199	.98506	5
56	.15529	.15719	6.3617	.98787	4	56	.17250	.17513	5.7101	.98501	4
57	.15557	.15749	6.3496	.98782	3	57	.17279	.17543	5.7004	.98496	3
58	.15586	.15779	6.3376	.98778	2	58	.17308	.17573	5.6906	.98491	2
59	.15615	.15809	6.3257	.98773	1	59	.17336	.17603	5.6809	.98486	1
60	.15643	.15838	6.3138	.98769	0	60	.17365	.17633	5.6713	.98481	0
′	Cos	Ctn	Tan	Sin	′	′	Cos	Ctn	Tan	Sin	′

TABLE OF NATURAL TRIGONOMETRIC FUNCTIONS (Cont.)

′	Sin	Tan	Ctn	Cos	′	′	Sin	Tan	Ctn	Cos	′
0	.17365	.17633	5.6713	.98481	60	0	.19081	.19438	5.1446	.98163	60
1	.17393	.17663	5.6617	.98476	59	1	.19109	.19468	5.1366	.98157	59
2	.17422	.17693	5.6521	.98471	58	2	.19138	.19498	5.1286	.98152	58
3	.17451	.17723	5.6425	.98466	57	3	.19167	.19529	5.1207	.98146	57
4	.17479	.17753	5.6329	.98461	56	4	.19195	.19559	5.1128	.98140	56
5	.17508	.17783	5.6234	.98455	55	5	.19224	.19589	5.1049	.98135	55
6	.17537	.17813	5.6140	.98450	54	6	.19252	.19619	5.0970	.98129	54
7	.17565	.17843	5.6045	.98445	53	7	.19281	.19649	5.0892	.98124	53
8	.17594	.17873	5.5951	.98440	52	8	.19309	.19680	5.0814	.98118	52
9	.17623	.17903	5.5857	.98435	51	9	.19338	.19710	5.0736	.98112	51
10	.17651	.17933	5.5764	.98430	50	10	.19366	.19740	5.0658	.98107	50
11	.17680	.17963	5.5671	.98425	49	11	.19395	.19770	5.0581	.98101	49
12	.17708	.17993	5.5578	.98420	48	12	.19423	.19801	5.0504	.98096	48
13	.17737	.18023	5.5485	.98414	47	13	.19452	.19831	5.0427	.98090	47
14	.17766	.18053	5.5393	.98409	46	14	.19481	.19861	5.0350	.98084	46
15	.17794	.18083	5.5301	.98404	45	15	.19509	.19891	5.0273	.98079	45
16	.17823	.18113	5.5209	.98399	44	16	.19538	.19921	5.0197	.98073	44
17	.17852	.18143	5.5118	.98394	43	17	.19566	.19952	5.0121	.98067	43
18	.17880	.18173	5.5026	.98389	42	18	.19595	.19982	5.0045	.98061	42
19	.17909	.18203	5.4936	.98383	41	19	.19623	.20012	4.9969	.98056	41
20	.17937	.18233	5.4845	.98378	40	20	.19652	.20042	4.9894	.98050	40
21	.17966	.18263	5.4755	.98373	39	21	.19680	.20073	4.9819	.98044	39
22	.17995	.18293	5.4665	.98368	38	22	.19709	.20103	4.9744	.98039	38
23	.18023	.18323	5.4575	.98362	37	23	.19737	.20133	4.9669	.98033	37
24	.18052	.18353	5.4486	.98357	36	24	.19766	.20164	4.9594	.98027	36
25	.18081	.18384	5.4397	.98352	35	25	.19794	.20194	4.9520	.98021	35
26	.18109	.18414	5.4308	.98347	34	26	.19823	.20224	4.9446	.98016	34
27	.18138	.18444	5.4219	.98341	33	27	.19851	.20254	4.9372	.98010	33
28	.18166	.18474	5.4131	.98336	32	28	.19880	.20285	4.9298	.98004	32
29	.18195	.18504	5.4043	.98331	31	29	.19908	.20315	4.9225	.97998	31
30	.18224	.18534	5.3955	.98325	30	30	.19937	.20345	4.9152	.97992	30
31	.18252	.18564	5.3868	.98320	29	31	.19965	.20376	4.9078	.97987	29
32	.18281	.18594	5.3781	.98315	28	32	.19994	.20406	4.9006	.97981	28
33	.18309	.18624	5.3694	.98310	27	33	.20022	.20436	4.8933	.97975	27
34	.18338	.18654	5.3607	.98304	26	34	.20051	.20466	4.8860	.97969	26
35	.18367	.18684	5.3521	.98299	25	35	.20079	.20497	4.8788	.97963	25
36	.18395	.18714	5.3435	.98294	24	36	.20108	.20527	4.8716	.97958	24
37	.18424	.18745	5.3349	.98288	23	37	.20136	.20557	4.8644	.97952	23
38	.18452	.18775	5.3263	.98283	22	38	.20165	.20588	4.8573	.97946	22
39	.18481	.18805	5.3178	.98277	21	39	.20193	.20618	4.8501	.97940	21
40	.18509	.18835	5.3093	.98272	20	40	.20222	.20648	4.8430	.97934	20
41	.18538	.18865	5.3008	.98267	19	41	.20250	.20679	4.8359	.97928	19
42	.18567	.18895	5.2924	.98261	18	42	.20279	.20709	4.8288	.97922	18
43	.18595	.18925	5.2839	.98256	17	43	.20307	.20739	4.8218	.97916	17
44	.18624	.18955	5.2755	.98250	16	44	.20336	.20770	4.8147	.97910	16
45	.18652	.18986	5.2672	.98245	15	45	.20364	.20800	4.8077	.97905	15
46	.18681	.19016	5.2588	.98240	14	46	.20393	.20830	4.8007	.97899	14
47	.18710	.19046	5.2505	.98234	13	47	.20421	.20861	4.7937	.97893	13
48	.18738	.19076	5.2422	.98229	12	48	.20450	.20891	4.7867	.97887	12
49	.18767	.19106	5.2339	.98223	11	49	.20478	.20921	4.7798	.97881	11
50	.18795	.19136	5.2257	.98218	10	50	.20507	.20952	4.7729	.97875	10
51	.18824	.19166	5.2174	.98212	9	51	.20535	.20982	4.7659	.97869	9
52	.18852	.19197	5.2092	.98207	8	52	.20563	.21013	4.7591	.97863	8
53	.18881	.19227	5.2011	.98201	7	53	.20592	.21043	4.7522	.97857	7
54	.18910	.19257	5.1929	.98196	6	54	.20620	.21073	4.7453	.97851	6
55	.18938	.19287	5.1848	.98190	5	55	.20649	.21104	4.7385	.97845	5
56	.18967	.19317	5.1767	.98185	4	56	.20677	.21134	4.7317	.97839	4
57	.18995	.19347	5.1686	.98179	3	57	.20706	.21164	4.7249	.97833	3
58	.19024	.19378	5.1606	.98174	2	58	.20734	.21195	4.7181	.97827	2
59	.19052	.19408	5.1526	.98168	1	59	.20763	.21225	4.7114	.97821	1
60	.19081	.19438	5.1446	.98163	0	60	.20791	.21256	4.7046	.97815	0
′	Cos	Ctn	Tan	Sin	′	′	Cos	Ctn	Tan	Sin	′

TABLE OF NATURAL TRIGONOMETRIC FUNCTIONS (Cont.)

′	Sin	Tan	Ctn	Cos	′	′	Sin	Tan	Ctn	Cos	′
0	.20791	.21256	4.7046	.97815	60	0	.22495	.23087	4.3315	.97437	60
1	.20820	.21286	4.6979	.97809	59	1	.22523	.23117	4.3257	.97430	59
2	.20848	.21316	4.6912	.97803	58	2	.22552	.23148	4.3200	.97424	58
3	.20877	.21347	4.6845	.97797	57	3	.22580	.23179	4.3143	.97417	57
4	.20905	.21377	4.6779	.97791	56	4	.22608	.23209	4.3086	.97411	56
5	.20933	.21408	4.6712	.97784	55	5	.22637	.23240	4.3029	.97404	55
6	.20962	.21438	4.6646	.97778	54	6	.22665	.23271	4.2972	.97398	54
7	.20990	.21469	4.6580	.97772	53	7	.22693	.23301	4.2916	.97391	53
8	.21019	.21499	4.6514	.97766	52	8	.22722	.23332	4.2859	.97384	52
9	.21047	.21529	4.6448	.97760	51	9	.22750	.23363	4.2803	.97378	51
10	.21076	.21560	4.6382	.97754	50	10	.22778	.23393	4.2747	.97371	50
11	.21104	.21590	4.6317	.97748	49	11	.22807	.23424	4.2691	.97365	49
12	.21132	.21621	4.6252	.97742	48	12	.22835	.23455	4.2635	.97358	48
13	.21161	.21651	4.6187	.97735	47	13	.22863	.23485	4.2580	.97351	47
14	.21189	.21682	4.6122	.97729	46	14	.22892	.23516	4.2524	.97345	46
15	.21218	.21712	4.6057	.97723	45	15	.22920	.23547	4.2468	.97338	45
16	.21246	.21743	4.5993	.97717	44	16	.22948	.23578	4.2413	.97331	44
17	.21275	.21773	4.5928	.97711	43	17	.22977	.23608	4.2358	.97325	43
18	.21303	.21804	4 5864	.97705	42	18	.23005	.23639	4.2303	.97318	42
19	.21331	.21834	4.5800	.97698	41	19	.23033	.23670	4.2248	.97311	41
20	.21360	.21864	4.5736	.97692	40	20	.23062	.23700	4.2193	.97304	40
21	.21388	.21895	4.5673	.97686	39	21	.23090	.23731	4.2139	.97298	39
22	.21417	.21925	4.5609	.97680	38	22	.23118	.23762	4.2084	.97291	38
23	.21445	.21956	4.5546	.97673	37	23	.23146	.23793	4.2030	.97284	37
24	.21474	.21986	4.5483	.97667	36	24	.23175	.23823	4.1976	.97278	36
25	.21502	.22017	4.5420	.97661	35	25	.23203	.23854	4.1922	.97271	35
26	.21530	.22047	4.5357	.97655	34	26	.23231	.23885	4.1868	.97264	34
27	.21559	.22078	4.5294	.97648	33	27	.23260	.23916	4.1814	.97257	33
28	.21587	.22108	4.5232	.97642	32	28	.23288	.23946	4.1760	.97251	32
29	.21616	.22139	4.5169	.97636	31	29	.23316	.23977	4.1706	.97244	31
30	.21644	.22169	4.5107	.97630	30	30	.23345	.24008	4.1653	.97237	30
31	.21672	.22200	4.5045	.97623	29	31	.23373	.24039	4.1600	.97230	29
32	.21701	.22231	4.4983	.97617	28	32	.23401	.24069	4.1547	.97223	28
33	.21729	.22261	4.4922	.97611	27	33	.23429	.24100	4.1493	.97217	27
34	.21758	.22292	4.4860	.97604	26	34	.23458	.24131	4.1441	.97210	26
35	.21786	.22322	4.4799	.97598	25	35	.23486	.24162	4.1388	.97203	25
36	.21814	.22353	4.4737	.97592	24	36	.23514	.24193	4.1335	.97196	24
37	.21843	.22383	4.4676	.97585	23	37	.23542	.24223	4.1282	.97189	23
38	.21871	.22414	4.4615	.97579	22	38	.23571	.24254	4.1230	.97182	22
39	.21899	.22444	4.4555	.97573	21	39	.23599	.24285	4.1178	.97176	21
40	.21928	.22475	4.4494	.97566	20	40	.23627	.24316	4.1126	.97169	20
41	.21956	.22505	4.4434	.97560	19	41	.23656	.24347	4.1074	.97162	19
42	.21985	.22536	4.4373	.97553	18	42	.23684	.24377	4.1022	.97155	18
43	.22013	.22567	4.4313	.97547	17	43	.23712	.24408	4.0970	.97148	17
44	.22041	.22597	4.4253	.97541	16	44	.23740	.24439	4.0918	.97141	16
45	.22070	.22628	4.4194	.97534	15	45	.23769	.24470	4.0867	.97134	15
46	.22098	.22658	4.4134	.97528	14	46	.23797	.24501	4.0815	.97127	14
47	.22126	.22689	4.4075	.97521	13	47	.23825	.24532	4.0764	.97120	13
48	.22155	.22719	4.4015	.97515	12	48	.23853	.24562	4.0713	.97113	12
49	.22183	.22750	4.3956	.97508	11	49	.23882	.24593	4.0662	.97106	11
50	.22212	.22781	4.3897	.97502	10	50	.23910	.24624	4.0611	.97100	10
51	.22240	.22811	4.3838	.97496	9	51	.23938	.24655	4.0560	.97093	9
52	.22268	.22842	4.3779	.97489	8	52	.23966	.24686	4.0509	.97086	8
53	.22297	.22872	4.3721	.97483	7	53	.23995	.24717	4.0459	.97079	7
54	.22325	.22903	4.3662	.97476	6	54	.24023	.24747	4.0408	.97072	6
55	.22353	.22934	4.3604	.97470	5	55	.24051	.24778	4.0358	.97065	5
56	.22382	.22964	4.3546	.97463	4	56	.24079	.24809	4.0308	.97058	4
57	.22410	.22995	4.3488	.97457	3	57	.24108	.24840	4.0257	.97051	3
58	.22438	.23026	4.3430	.97450	2	58	.24136	.24871	4.0207	.97044	2
59	.22467	.23056	4.3372	.97444	1	59	.24164	.24902	4.0158	.97037	1
60	.22495	.23087	4.3315	.97437	0	60	.24192	.24933	4.0108	.97030	0
′	Cos	Ctn	Tan	Sin	′	′	Cos	Ctn	Tan	Sin	′

14°				165°		15°				164°
′	Sin	Tan	Ctn	Cos	′	′	Sin	Tan	Ctn	Cos	′

′	Sin	Tan	Ctn	Cos	′	′	Sin	Tan	Ctn	Cos	′
0	.24192	.24933	4.0108	.97030	60	0	.25882	.26795	3.7321	.96593	60
1	.24220	.24964	4.0058	.97023	59	1	.25910	.26826	3.7277	.96585	59
2	.24249	.24995	4.0009	.97015	58	2	.25938	.26857	3.7234	.96578	58
3	.24277	.25026	3.9959	.97008	57	3	.25966	.26888	3.7191	.96570	57
4	.24305	.25056	3.9910	.97001	56	4	.25994	.26920	3.7148	.96562	56
5	.24333	.25087	3.9861	.96994	55	5	.26022	.26951	3.7105	.96555	55
6	.24362	.25118	3.9812	.96987	54	6	.26050	.26982	3.7062	.96547	54
7	.24390	.25149	3.9763	.96980	53	7	.26079	.27013	3.7019	.96540	53
8	.24418	.25180	3.9714	.96973	52	8	.26107	.27044	3.6976	.96532	52
9	.24446	.25211	3.9665	.96966	51	9	.26135	.27076	3.6933	.96524	51
10	.24474	.25242	3.9617	.96959	50	10	.26163	.27107	3.6891	.96517	50
11	.24503	.25273	3.9568	.96952	49	11	.26191	.27138	3.6848	.96509	49
12	.24531	.25304	3.9520	.96945	48	12	.26219	.27169	3.6806	.96502	48
13	.24559	.25335	3.9471	.96937	47	13	.26247	.27201	3.6764	.96494	47
14	.24587	.25366	3.9423	.96930	46	14	.26275	.27232	3.6722	.96486	46
15	.24615	.25397	3.9375	.96923	45	15	.26303	.27263	3.6680	.96479	45
16	.24644	.25428	3.9327	.96916	44	16	.26331	.27294	3.6638	.96471	44
17	.24672	.25459	3.9279	.96909	43	17	.26359	.27326	3.6596	.96463	43
18	.24700	.25490	3.9232	.96902	42	18	.26387	.27357	3.6554	.96456	42
19	.24728	.25521	3.9184	.96894	41	19	.26415	.27388	3.6512	.96448	41
20	.24756	.25552	3.9136	.96887	40	20	.26443	.27419	3.6470	.96440	40
21	.24784	.25583	3.9089	.96880	39	21	.26471	.27451	3.6429	.96433	39
22	.24813	.25614	3.9042	.96873	38	22	.26500	.27482	3.6387	.96425	38
23	.24841	.25645	3.8995	.96866	37	23	.26528	.27513	3.6346	.96417	37
24	.24869	.25676	3.8947	.96858	36	24	.26556	.27545	3.6305	.96410	36
25	.24897	.25707	3.8900	.96851	35	25	.26584	.27576	3.6264	.96402	35
26	.24925	.25738	3.8854	.96844	34	26	.26612	.27607	3.6222	.96394	34
27	.24954	.25769	3.8807	.96837	33	27	.26640	.27638	3.6181	.96386	33
28	.24982	.25800	3.8760	.96829	32	28	.26668	.27670	3.6140	.96379	32
29	.25010	.25831	3.8714	.96822	31	29	.26696	.27701	3.6100	.96371	31
30	.25038	.25862	3.8667	.96815	30	30	.26724	.27732	3.6059	.96363	30
31	.25066	.25893	3.8621	.96807	29	31	.26752	.27764	3.6018	.96355	29
32	.25094	.25924	3.8575	.96800	28	32	.26780	.27795	3.5978	.96347	28
33	.25122	.25955	3.8528	.96793	27	33	.26808	.27826	3.5937	.96340	27
34	.25151	.25986	3.8482	.96786	26	34	.26836	.27858	3.5897	.96332	26
35	.25179	.26017	3.8436	.96778	25	35	.26864	.27889	3.5856	.96324	25
36	.25207	.26048	3.8391	.96771	24	36	.26892	.27921	3.5816	.96316	24
37	.25235	.26079	3.8345	.96764	23	37	.26920	.27952	3.5776	.96308	23
38	.25263	.26110	3.8299	.96756	22	38	.26948	.27983	3.5736	.96301	22
39	.25291	.26141	3.8254	.96749	21	39	.26976	.28015	3.5696	.96293	21
40	.25320	.26172	3.8208	.96742	20	40	.27004	.28046	3.5656	.96285	20
41	.25348	.26203	3.8163	.96734	19	41	.27032	.28077	3.5616	.96277	19
42	.25376	.26235	3.8118	.96727	18	42	.27060	.28109	3.5576	.96269	18
43	.25404	.26266	3.8073	.96719	17	43	.27088	.28140	3.5536	.96261	17
44	.25432	.26297	3.8028	.96712	16	44	.27116	.28172	3.5497	.96253	16
45	.25460	.26328	3.7983	.96705	15	45	.27144	.28203	3.5457	.96246	15
46	.25488	.26359	3.7938	.96697	14	46	.27172	.28234	3.5418	.96238	14
47	.25516	.26390	3.7893	.96690	13	47	.27200	.28266	3.5379	.96230	13
48	.25545	.26421	3.7848	.96682	12	48	.27228	.28297	3.5339	.96222	12
49	.25573	.26452	3.7804	.96675	11	49	.27256	.28329	3.5300	.96214	11
50	.25601	.26483	3.7760	.96667	10	50	.27284	.28360	3.5261	.96206	10
51	.25629	.26515	3.7715	.96660	9	51	.27312	.28391	3.5222	.96198	9
52	.25657	.26546	3.7671	.96653	8	52	.27340	.28423	3.5183	.96190	8
53	.25685	.26577	3.7627	.96645	7	53	.27368	.28454	3.5144	.96182	7
54	.25713	.26308	3.7583	.96638	6	54	.27396	.28486	3.5105	.96174	6
55	.25741	.26639	3.7539	.96630	5	55	.27424	.28517	3.5067	.96166	5
56	.25769	.26670	3.7495	.96623	4	56	.27452	.28549	3.5028	.96158	4
57	.25798	.26701	3.7451	.96615	3	57	.27480	.28580	3.4989	.96150	3
58	.25826	.26733	3.7408	.96608	2	58	.27508	.28612	3.4951	.96142	2
59	.25854	.26764	3.7364	.96600	1	59	.27536	.28643	3.4912	.96134	1
60	.25882	.26795	3.7321	.96593	0	60	.27564	.28675	3.4874	.96126	0
′	Cos	Ctn	Tan	Sin	′	′	Cos	Ctn	Tan	Sin	′

TABLE OF NATURAL TRIGONOMETRIC FUNCTIONS (Cont.)

16°   **163°**   **17°**   **162°**

′	Sin	Tan	Ctn	Cos	′		′	Sin	Tan	Ctn	Cos	′
0	.27564	.28675	3.4874	.96126	60		0	.29237	.30573	3.2709	.95630	60
1	.27592	.28706	3.4836	.96118	59		1	.29265	.30605	3.2675	.95622	59
2	.27620	.28738	3.4798	.96110	58		2	.29293	.30637	3.2641	.95613	58
3	.27648	.28769	3.4760	.96102	57		3	.29321	.30669	3.2607	.95605	57
4	.27676	.28801	3.4722	.96094	56		4	.29348	.30700	3.2573	.95596	56
5	.27704	.28832	3.4684	.96086	55		5	.29376	.30732	3.2539	.95588	55
6	.27731	.28864	3.4646	.96078	54		6	.29404	.30764	3.2506	.95579	54
7	.27759	.28895	3.4608	.96070	53		7	.29432	.30796	3.2472	.95571	53
8	.27787	.28927	3.4570	.96062	52		8	.29460	.30828	3.2438	.95562	52
9	.27815	.28958	3.4533	.96054	51		9	.29487	.30860	3.2405	.95554	51
10	.27843	.28990	3.4495	.96046	50		10	.29515	.30891	3.2371	.95545	50
11	.27871	.29021	3.4458	.96037	49		11	.29543	.30923	3.2338	.95536	49
12	.27899	.29053	3.4420	.96029	48		12	.29571	.30955	3.2305	.95528	48
13	.27927	.29084	3.4383	.96021	47		13	.29599	.30987	3.2272	.95519	47
14	.27955	.29116	3.4346	.96013	46		14	.29626	.31019	3.2238	.95511	46
15	.27983	.29147	3.4308	.96005	45		15	.29654	.31051	3.2205	.95502	45
16	.28011	.29179	3.4271	.95997	44		16	.29682	.31083	3.2172	.95493	44
17	.28039	.29210	3.4234	.95989	43		17	.29710	.31115	3.2139	.95485	43
18	.28067	.29242	3.4197	.95981	42		18	.29737	.31147	3.2106	.95476	42
19	.28095	.29274	3.4160	.95972	41		19	.29765	.31178	3.2073	.95467	41
20	.28123	.29305	3.4124	.95964	40		20	.29793	.31210	3.2041	.95459	40
21	.28150	.29337	3.4087	.95956	39		21	.29821	.31242	3.2008	.95450	39
22	.28178	.29368	3.4050	.95948	38		22	.29849	.31274	3.1975	.95441	38
23	.28206	.29400	3.4014	.95940	37		23	.29876	.31306	3.1943	.95433	37
24	.28234	.29432	3.3977	.95931	36		24	.29904	.31338	3.1910	.95424	36
25	.28262	.29463	3.3941	.95923	35		25	.29932	.31370	3.1878	.95415	35
26	.28290	.29495	3.3904	.95915	34		26	.29960	.31402	3.1845	.95407	34
27	.28318	.29526	3.3868	.95907	33		27	.29987	.31434	3.1813	.95398	33
28	.28346	.29558	3.3832	.95898	32		28	.30015	.31466	3.1780	.95389	32
29	.28374	.29590	3.3796	.95890	31		29	.30043	.31498	3.1748	.95380	31
30	.28402	.29621	3.3759	.95882	30		30	.30071	.31530	3.1716	.95372	30
31	.28429	.29653	3.3723	.95874	29		31	.30098	.31562	3.1684	.95363	29
32	.28457	.29685	3.3687	.95865	28		32	.30126	.31594	3.1652	.95354	28
33	.28485	.29716	3.3652	.95857	27		33	.30154	.31626	3.1620	.95345	27
34	.28513	.29748	3.3616	.95849	26		34	.30182	.31658	3.1588	.95337	26
35	.28541	.29780	3.3580	.95841	25		35	.30209	.31690	3.1556	.95328	25
36	.28569	.29811	3.3544	.95832	24		36	.30237	.31722	3.1524	.95319	24
37	.28597	.29843	3.3509	.95824	23		37	.30265	.31754	3.1492	.95310	23
38	.28625	.29875	3.3473	.95816	22		38	.30292	.31786	3.1460	.95301	22
39	.28652	.29906	3.3438	.95807	21		39	.30320	.31818	3.1429	.95293	21
40	.28680	.29938	3.3402	.95799	20		40	.30348	.31850	3.1397	.95284	20
41	.28708	.29970	3.3367	.95791	19		41	.30376	.31882	3.1366	.95275	19
42	.28736	.30001	3.3332	.95782	18		42	.30403	.31914	3.1334	.95266	18
43	.28764	.30033	3.3297	.95774	17		43	.30431	.31946	3.1303	.95257	17
44	.28792	.30065	3.3261	.95766	16		44	.30459	.31978	3.1271	.95248	16
45	.28820	.30097	3.3226	.95757	15		45	.30486	.32010	3.1240	.95240	15
46	.28847	.30128	3.3191	.95749	14		46	.30514	.32042	3.1209	.95231	14
47	.28875	.30160	3.3156	.95740	13		47	.30542	.32074	3.1178	.95222	13
48	.28903	.30192	3.3122	.95732	12		48	.30570	.32106	3.1146	.95213	12
49	.28931	.30224	3.3087	.95724	11		49	.30597	.32139	3.1115	.95204	11
50	.28959	.30255	3.3052	.95715	10		50	.30625	.32171	3.1084	.95195	10
51	.28987	.30287	3.3017	.95707	9		51	.30653	.32203	3.1053	.95186	9
52	.29015	.30319	3.2983	.95698	8		52	.30630	.32235	3.1022	.95177	8
53	.29042	.30351	3.2948	.95690	7		53	.30708	.32267	3.0991	.95168	7
54	.29070	.30382	3.2914	.95681	6		54	.30736	.32299	3.0961	.95159	6
55	.29098	.30414	3.2879	.95673	5		55	.30763	.32331	3.0930	.95150	5
56	.29126	.30446	3.2845	.95664	4		56	.30791	.32363	3.0899	.95142	4
57	.29154	.30478	3.2811	.95656	3		57	.30819	.32396	3.0868	.95133	3
58	.29182	.30509	3.2777	.95647	2		58	.30846	.32428	3.0838	.95124	2
59	.29209	.30541	3.2743	.95639	1		59	.30874	.32460	3.0807	.95115	1
60	.29237	.30573	3.2709	.95630	0		60	.30902	.32492	3.0777	.95106	0
′	Cos	Ctn	Tan	Sin	′		′	Cos	Ctn	Tan	Sin	′

106°   **73°**   **107°**   **72°**

18°   **161°**   **19°**   **160°**

′	Sin	Tan	Ctn	Cos	′		′	Sin	Tan	Ctn	Cos	′
0	.30902	.32492	3.0777	.95106	60		0	.32557	.34433	2.9042	.94552	60
1	.30929	.32524	3.0746	.95097	59		1	.32584	.34465	2.9015	.94542	59
2	.30957	.32556	3.0716	.95088	58		2	.32612	.34498	2.8987	.94533	58
3	.30985	.32588	3.0686	.95079	57		3	.32639	.34530	2.8960	.94523	57
4	.31012	.32621	3.0655	.95070	56		4	.32667	.34563	2.8933	.94514	56
5	.31040	.32653	3.0625	.95061	55		5	.32694	.34596	2.8905	.94504	55
6	.31068	.32685	3.0595	.95052	54		6	.32722	.34628	2.8878	.94495	54
7	.31095	.32717	3.0565	.95043	53		7	.32749	.34661	2.8851	.94485	53
8	.31123	.32749	3.0535	.95033	52		8	.32777	.34693	2.8824	.94476	52
9	.31151	.32782	3.0505	.95024	51		9	.32804	.34726	2.8797	.94466	51
10	.31178	.32814	3.0475	.95015	50		10	.32832	.34758	2.8770	.94457	50
11	.31206	.32846	3.0445	.95006	49		11	.32859	.34791	2.8743	.94447	49
12	.31233	.32878	3.0415	.94997	48		12	.32887	.34824	2.8716	.94438	48
13	.31261	.32911	3.0385	.94988	47		13	.32914	.34856	2.8689	.94428	47
14	.31289	.32943	3.0356	.94979	46		14	.32942	.34889	2.8662	.94418	46
15	.31316	.32975	3.0326	.94970	45		15	.32969	.34922	2.8636	.94409	45
16	.31344	.33007	3.0296	.94961	44		16	.32997	.34954	2.8609	.94399	44
17	.31372	.33040	3.0267	.94952	43		17	.33024	.34987	2.8582	.94390	43
18	.31399	.33072	3.0237	.94943	42		18	.33051	.35020	2.8556	.94380	42
19	.31427	.33104	3.0208	.94933	41		19	.33079	.35052	2.8529	.94370	41
20	.31454	.33136	3.0178	.94924	40		20	.33106	.35085	2.8502	.94361	40
21	.31482	.33169	3.0149	.94915	39		21	.33134	.35118	2.8476	.94351	39
22	.31510	.33201	3.0120	.94906	38		22	.33161	.35150	2.8449	.94342	38
23	.31537	.33233	3.0090	.94897	37		23	.33189	.35183	2.8423	.94332	37
24	.31565	.33266	3.0061	.94888	36		24	.33216	.35216	2.8397	.94322	36
25	.31593	.33298	3.0032	.94878	35		25	.33244	.35248	2.8370	.94313	35
26	.31620	.33330	3.0003	.94869	34		26	.33271	.35281	2.8344	.94303	34
27	.31648	.33363	2.9974	.94860	33		27	.33298	.35314	2.8318	.94293	33
28	.31675	.33395	2.9945	.94851	32		28	.33326	.35346	2.8291	.94284	32
29	.31703	.33427	2.9916	.94842	31		29	.33353	.35379	2.8265	.94274	31
30	.31730	.33460	2.9887	.94832	30		30	.33381	.35412	2.8239	.94264	30
31	.31758	.33492	2.9858	.94823	29		31	.33408	.35445	2.8213	.94254	29
32	.31786	.33524	2.9829	.94814	28		32	.33436	.35477	2.8187	.94245	28
33	.31813	.33557	2.9800	.94805	27		33	.33463	.35510	2.8161	.94235	27
34	.31841	.33589	2.9772	.94795	26		34	.33490	.35543	2.8135	.94225	26
35	.31868	.33621	2.9743	.94786	25		35	.33518	.35576	2.8109	.94215	25
36	.31896	.33654	2.9714	.94777	24		36	.33545	.35608	2.8083	.94206	24
37	.31923	.33686	2.9686	.94768	23		37	.33573	.35641	2.8057	.94196	23
38	.31951	.33718	2.9657	.94758	22		38	.33600	.35674	2.8032	.94186	22
39	.31979	.33751	2.9629	.94749	21		39	.33627	.35707	2.8006	.94176	21
40	.32006	.33783	2.9600	.94740	20		40	.33655	.35740	2.7980	.94167	20
41	.32034	.33816	2.9572	.94730	19		41	.33682	.35772	2.7955	.94157	19
42	.32061	.33848	2.9544	.94721	18		42	.33710	.35805	2.7929	.94147	18
43	.32089	.33881	2.9515	.94712	17		43	.33737	.35838	2.7903	.94137	17
44	.32116	.33913	2.9487	.94702	16		44	.33764	.35871	2.7878	.94127	16
45	.32144	.33945	2.9459	.94693	15		45	.33792	.35904	2.7852	.94118	15
46	.32171	.33978	2.9431	.94684	14		46	.33819	.35937	2.7827	.94108	14
47	.32199	.34010	2.9403	.94674	13		47	.33846	.35969	2.7801	.94098	13
48	.32227	.34043	2.9375	.94665	12		48	.33874	.36002	2.7776	.94088	12
49	.32254	.34075	2.9347	.94656	11		49	.33901	.36035	2.7751	.94078	11
50	.32282	.34108	2.9319	.94646	10		50	.33929	.36068	2.7725	.94068	10
51	.32309	.34140	2.9291	.94637	9		51	.33956	.36101	2.7700	.94058	9
52	.32337	.34173	2.9263	.94627	8		52	.33983	.36134	2.7675	.94049	8
53	.32364	.34205	2.9235	.94618	7		53	.34011	.36167	2.7650	.94039	7
54	.32392	.34238	2.9208	.94609	6		54	.34038	.36199	2.7625	.94029	6
55	.32419	.34270	2.9180	.94599	5		55	.34065	.36232	2.7600	.94019	5
56	.32447	.34303	2.9152	.94590	4		56	.34093	.36265	2.7575	.94009	4
57	.32474	.34335	2.9125	.94580	3		57	.34120	.36298	2.7550	.93999	3
58	.32502	.34368	2.9097	.94571	2		58	.34147	.36331	2.7525	.93989	2
59	.32529	.34400	2.9070	.94561	1		59	.34175	.36364	2.7500	.93979	1
60	.32557	.34433	2.9042	.94552	0		60	.34202	.36397	2.7475	.93969	0
′	Cos	Ctn	Tan	Sin	′		′	Cos	Ctn	Tan	Sin	′

108°   **71°**   **109°**   **70°**

20°				159°		21°				158°
′	Sin	Tan	Ctn	Cos	′	′	Sin	Tan	Ctn	Cos	′
0	.34202	.36397	2.7475	.93969	60	0	.35837	.38386	2.6051	.93358	60
1	.34229	.36430	2.7450	.93959	59	1	.35864	.38420	2.6028	.93348	59
2	.34257	.36463	2.7425	.93949	58	2	.35891	.38453	2.6006	.93337	58
3	.34284	.36496	2.7400	.93939	57	3	.35918	.38487	2.5983	.93327	57
4	.34311	.36529	2.7376	.93929	56	4	.35945	.38520	2.5961	.93316	56
5	.34339	.36562	2.7351	.93919	55	5	.35973	.38553	2.5938	.93306	55
6	.34366	.36595	2.7326	.93909	54	6	.36000	.38587	2.5916	.93295	54
7	.34393	.36628	2.7302	.93899	53	7	.36027	.38620	2.5893	.93285	53
8	.34421	.36661	2.7277	.93889	52	8	.36054	.38654	2.5871	.93274	52
9	.34448	.36694	2.7253	.93879	51	9	.36081	.38687	2.5848	.93264	51
10	.34475	.36727	2.7228	.93869	50	10	.36108	.38721	2.5826	.93253	50
11	.34503	.36760	2.7204	.93859	49	11	.36135	.38754	2.5804	.93243	49
12	.34530	.36793	2.7179	.93849	48	12	.36162	.38787	2.5782	.93232	48
13	.34557	.36826	2.7155	.93839	47	13	.36190	.38821	2.5759	.93222	47
14	.34584	.36859	2.7130	.93829	46	14	.36217	.38854	2.5737	.93211	46
15	.34612	.36892	2.7106	.93819	45	15	.36244	.38888	2.5715	.93201	45
16	.34639	.36925	2.7082	.93809	44	16	.36271	.38921	2.5693	.93190	44
17	.34666	.36958	2.7058	.93799	43	17	.36298	.38955	2.5671	.93180	43
18	.34694	.36991	2.7034	.93789	42	18	.36325	.38988	2.5649	.93169	42
19	.34721	.37024	2.7009	.93779	41	19	.36352	.39022	2.5627	.93159	41
20	.34748	.37057	2.6985	.93769	40	20	.36379	.39055	2.5605	.93148	40
21	.34775	.37090	2.6961	.93759	39	21	.36406	.39089	2.5583	.93137	39
22	.34803	.37123	2.6937	.93748	38	22	.36434	.39122	2.5561	.93127	38
23	.34830	.37157	2.6913	.93738	37	23	.36461	.39156	2.5539	.93116	37
24	.34857	.37190	2.6889	.93728	36	24	.36488	.39190	2.5517	.93106	36
25	.34884	.37223	2.6865	.93718	35	25	.36515	.39223	2.5495	.93095	35
26	.34912	.37256	2.6841	.93708	34	26	.36542	.39257	2.5473	.93084	34
27	.34939	.37289	2.6818	.93698	33	27	.36569	.39290	2.5452	.93074	33
28	.34966	.37322	2.6794	.93688	32	28	.36596	.39324	2.5430	.93063	32
29	.34993	.37355	2.6770	.93677	31	29	.36623	.39357	2.5408	.93052	31
30	.35021	.37388	2.6746	.93667	30	30	.36650	.39391	2.5386	.93042	30
31	.35048	.37422	2.6723	.93657	29	31	.36677	.39425	2.5365	.93031	29
32	.35075	.37455	2.6699	.93647	28	32	.36704	.39458	2.5343	.93020	28
33	.35102	.37488	2.6675	.93637	27	33	.36731	.39492	2.5322	.93010	27
34	.35130	.37521	2.6652	.93626	26	34	.36758	.39526	2.5300	.92999	26
35	.35157	.37554	2.6628	.93616	25	35	.36785	.39559	2.5279	.92988	25
36	.35184	.37588	2.6605	.93606	24	36	.36812	.39593	2.5257	.92978	24
37	.35211	.37621	2.6581	.93596	23	37	.36839	.39626	2.5236	.92967	23
38	.35239	.37654	2.6558	.93585	22	38	.36867	.39660	2.5214	.92956	22
39	.35266	.37687	2.6534	.93575	21	39	.36894	.39694	2.5193	.92945	21
40	.35293	.37720	2.6511	.93565	20	40	.36921	.39727	2.5172	.92935	20
41	.35320	.37754	2.6488	.93555	19	41	.36948	.39761	2.5150	.92924	19
42	.35347	.37787	2.6464	.93544	18	42	.36975	.39795	2.5129	.92913	18
43	.35375	.37820	2.6441	.93534	17	43	.37002	.39829	2.5108	.92902	17
44	.35402	.37853	2.6418	.93524	16	44	.37029	.39862	2.5086	.92892	16
45	.35429	.37887	2.6395	.93514	15	45	.37056	.39896	2.5065	.92881	15
46	.35456	.37920	2.6371	.93503	14	46	.37083	.39930	2.5044	.92870	14
47	.35484	.37953	2.6348	.93493	13	47	.37110	.39963	2.5023	.92859	13
48	.35511	.37986	2.6325	.93483	12	48	.37137	.39997	2.5002	.92849	12
49	.35538	.38020	2.6302	.93472	11	49	.37164	.40031	2.4981	.92838	11
50	.35565	.38053	2.6279	.93462	10	50	.37191	.40065	2.4960	.92827	10
51	.35592	.38086	2.6256	.93452	9	51	.37218	.40098	2.4939	.92816	9
52	.35619	.38120	2.6233	.93441	8	52	.37245	.40132	2.4918	.92805	8
53	.35647	.38153	2.6210	.93431	7	53	.37272	.40166	2.4897	.92794	7
54	.35674	.38186	2.6187	.93420	6	54	.37299	.40200	2.4876	.92784	6
55	.35701	.38220	2.6165	.93410	5	55	.37326	.40234	2.4855	.92773	5
56	.35728	.38253	2.6142	.93400	4	56	.37353	.40267	2.4834	.92762	4
57	.35755	.38286	2.6119	.93389	3	57	.37380	.40301	2.4813	.92751	3
58	.35782	.38320	2.6096	.93379	2	58	.37407	.40335	2.4792	.92740	2
59	.35810	.38353	2.6074	.93368	1	59	.37434	.40369	2.4772	.92729	1
60	.35837	.38386	2.6051	.93358	0	60	.37461	.40403	2.4751	.92718	0
′	Cos	Ctn	Tan	Sin	′	′	Cos	Ctn	Tan	Sin	′

110°				69°		111°				68°

266

22°				157°		23°				156°
′	Sin	Tan	Ctn	Cos	′	′	Sin	Tan	Ctn	Cos	′
0	.37461	.40403	2.4751	.92718	60	0	.39073	.42447	2.3559	.92050	60
1	.37488	.40436	2.4730	.92707	59	1	.39100	.42482	2.3539	.92039	59
2	.37515	.40470	2.4709	.92697	58	2	.39127	.42516	2.3520	.92028	58
3	.37542	.40504	2.4689	.92686	57	3	.39153	.42551	2.3501	.92016	57
4	.37569	.40538	2.4668	.92675	56	4	.39180	.42585	2.3483	.92005	56
5	.37595	.40572	2.4648	.92664	55	5	.39207	.42619	2.3464	.91994	55
6	.37622	.40606	2.4627	.92653	54	6	.39234	.42654	2.3445	.91982	54
7	.37649	.40640	2.4606	.92642	53	7	.39260	.42688	2.3426	.91971	53
8	.37676	.40674	2.4586	.92631	52	8	.39287	.42722	2.3407	.91959	52
9	.37703	.40707	2.4566	.92620	51	9	.39314	.42757	2.3388	.91948	51
10	.37730	.40741	2.4545	.92609	50	10	.39341	.42791	2.3369	.91936	50
11	.37757	.40775	2.4525	.92598	49	11	.39367	.42826	2.3351	.91925	49
12	.37784	.40809	2.4504	.92587	48	12	.39394	.42860	2.3332	.91914	48
13	.37811	.40843	2.4484	.92576	47	13	.39421	.42894	2.3313	.91902	47
14	.37838	.40877	2.4464	.92565	46	14	.39448	.42929	2.3294	.91891	46
15	.37865	.40911	2.4443	.92554	45	15	.39474	.42963	2.3276	.91879	45
16	.37892	.40945	2.4423	.92543	44	16	.39501	.42998	2.3257	.91868	44
17	.37919	.40979	2.4403	.92532	43	17	.39528	.43032	2.3238	.91856	43
18	.37946	.41013	2.4383	.92521	42	18	.39555	.43067	2.3220	.91845	42
19	.37973	.41047	2.4362	.92510	41	19	.39581	.43101	2.3201	.91833	41
20	.37999	.41081	2.4342	.92499	40	20	.39608	.43136	2.3183	.91822	40
21	.38026	.41115	2.4322	.92488	39	21	.39635	.43170	2.3164	.91810	39
22	.38053	.41149	2.4302	.92477	38	22	.39661	.43205	2.3146	.91799	38
23	.38080	.41183	2.4282	.92466	37	23	.39688	.43239	2.3127	.91787	37
24	.38107	.41217	2.4262	.92455	36	24	.39715	.43274	2.3109	.91775	36
25	.38134	.41251	2.4242	.92444	35	25	.39741	.43308	2.3090	.91764	35
26	.38161	.41285	2.4222	.92432	34	26	.39768	.43343	2.3072	.91752	34
27	.38188	.41319	2.4202	.92421	33	27	.39795	.43378	2.3053	.91741	33
28	.38215	.41353	2.4182	.92410	32	28	.39822	.43412	2.3035	.91729	32
29	.38241	.41387	2.4162	.92399	31	29	.39848	.43447	2.3017	.91718	31
30	.38268	.41421	2.4142	.92388	30	30	.39875	.43481	2.2998	.91706	30
31	.38295	.41455	2.4122	.92377	29	31	.39902	.43516	2.2980	.91694	29
32	.38322	.41490	2.4102	.92366	28	32	.39928	.43550	2.2962	.91683	28
33	.38349	.41524	2.4083	.92355	27	33	.39955	.43585	2.2944	.91671	27
34	.38376	.41558	2.4063	.92343	26	34	.39982	.43620	2.2925	.91660	26
35	.38403	.41592	2.4043	.92332	25	35	.40008	.43654	2.2907	.91648	25
36	.38430	.41626	2.4023	.92321	24	36	.40035	.43689	2.2889	.91636	24
37	.38456	.41660	2.4004	.92310	23	37	.40062	.43724	2.2871	.91625	23
38	.38483	.41694	2.3984	.92299	22	38	.40088	.43758	2.2853	.91613	22
39	.38510	.41728	2.3964	.92287	21	39	.40115	.43793	2.2835	.91601	21
40	.38537	.41763	2.3945	.92276	20	40	.40141	.43828	2.2817	.91590	20
41	.38564	.41797	2.3925	.92265	19	41	.40168	.43862	2.2799	.91578	19
42	.38591	.41831	2.3906	.92254	18	42	.40195	.43897	2.2781	.91566	18
43	.38617	.41865	2.3886	.92243	17	43	.40221	.43932	2.2763	.91555	17
44	.38644	.41899	2.3867	.92231	16	44	.40248	.43966	2.2745	.91543	16
45	.38671	.41933	2.3847	.92220	15	45	.40275	.44001	2.2727	.91531	15
46	.38698	.41968	2.3828	.92209	14	46	.40301	.44036	2.2709	.91519	14
47	.38725	.42002	2.3808	.92198	13	47	.40328	.44071	2.2691	.91508	13
48	.38752	.42036	2.3789	.92186	12	48	.40355	.44105	2.2673	.91496	12
49	.38778	.42070	2.3770	.92175	11	49	.40381	.44140	2.2655	.91484	11
50	.38805	.42105	2.3750	.92164	10	50	.40408	.44175	2.2637	.91472	10
51	.38832	.42139	2.3731	.92152	9	51	.40434	.44210	2.2620	.91461	9
52	.38859	.42173	2.3712	.92141	8	52	.40461	.44244	2.2602	.91449	8
53	.38886	.42207	2.3693	.92130	7	53	.40488	.44279	2.2584	.91437	7
54	.38912	.42242	2.3673	.92119	6	54	.40514	.44314	2.2566	.91425	6
55	.38939	.42276	2.3654	.92107	5	55	.40541	.44349	2.2549	.91414	5
56	.38966	.42310	2.3635	.92096	4	56	.40567	.44384	2.2531	.91402	4
57	.38993	.42345	2.3616	.92085	3	57	.40594	.44418	2.2513	.91390	3
58	.39020	.42379	2.3597	.92073	2	58	.40621	.44453	2.2496	.91378	2
59	.39046	.42413	2.3578	.92062	1	59	.40647	.44488	2.2478	.91366	1
60	.39073	.42447	2.3559	.92050	0	60	.40674	.44523	2.2460	.91355	0
′	Cos	Ctn	Tan	Sin	′	′	Cos	Ctn	Tan	Sin	′

112°				67°		113°				66°

267

24°   **155°**   **25°**   **154°**

′	Sin	Tan	Ctn	Cos	′		′	Sin	Tan	Ctn	Cos	′
0	.40674	.44523	2.2460	.91355	60		0	.42262	.46631	2.1445	.90631	60
1	.40700	.44558	2.2443	.91343	59		1	.42288	.46666	2.1429	.90618	59
2	.40727	.44593	2.2425	.91331	58		2	.42315	.46702	2.1413	.90606	58
3	.40753	.44627	2.2408	.91319	57		3	.42341	.46737	2.1396	.90594	57
4	.40780	.44662	2.2390	.91307	56		4	.42367	.46772	2.1380	.90582	56
5	.40806	.44697	2.2373	.91295	55		5	.42394	.46808	2.1364	.90569	55
6	.40833	.44732	2.2355	.91283	54		6	.42420	.46843	2.1348	.90557	54
7	.40860	.44767	2.2338	.91272	53		7	.42446	.46879	2.1332	.90545	53
8	.40886	.44802	2.2320	.91260	52		8	.42473	.46914	2.1315	.90532	52
9	.40913	.44837	2.2303	.91248	51		9	.42499	.46950	2.1299	.90520	51
10	.40939	.44872	2.2286	.91236	50		10	.42525	.46985	2.1283	.90507	50
11	.40966	.44907	2.2268	.91224	49		11	.42552	.47021	2.1267	.90495	49
12	.40992	.44942	2.2251	.91212	48		12	.42578	.47056	2.1251	90483	48
13	.41019	.44977	2.2234	.91200	47		13	.42604	.47092	2.1235	.90470	47
14	.41045	.45012	2.2216	.91188	46		14	.42631	.47128	2.1219	.90458	46
15	.41072	.45047	2.2199	.91176	45		15	.42657	.47163	2 1203	.90446	45
16	.41098	.45082	2.2182	.91164	44		16	.42683	.47199	2.1187	.90433	44
17	.41125	.45117	2.2165	.91152	43		17	.42709	.47234	2.1171	.90421	43
18	.41151	.45152	2.2148	.91140	42		18	.42736	.47270	2.1155	.90408	42
19	.41178	.45187	2.2130	.91128	41		19	.42762	.47305	2.1139	.90396	41
20	.41204	.45222	2.2113	.91116	40		20	.42788	.47341	2.1123	.90383	40
21	.41231	.45257	2.2096	.91104	39		21	.42815	.47377	2.1107	.90371	39
22	.41257	.45292	2.2079	.91092	38		22	.42841	.47412	2.1092	.90358	38
23	.41284	.45327	2.2062	.91080	37		23	.42867	.47448	2.1076	.90346	37
24	.41310	.45362	2.2045	.91068	36		24	.42894	.47483	2.1060	.90334	36
25	.41337	.45397	2.2028	.91056	35		25	.42920	.47519	2.1044	.90321	35
26	.41363	.45432	2.2011	.91044	34		26	.42946	.47555	2.1028	.90309	34
27	.41390	.45467	2.1994	.91032	33		27	.42972	.47590	2.1013	.90296	33
28	.41416	.45502	2.1977	.91020	32		28	.42999	.47626	2.0997	.90284	32
29	.41443	.45538	2.1960	.91008	31		29	.43025	.47662	2.0981	.90271	31
30	.41469	.45573	2.1943	.90996	30		30	.43051	.47698	2.0965	.90259	30
31	.41496	.45608	2.1926	.90984	29		31	.43077	.47733	2.0950	.90246	29
32	.41522	.45643	2.1909	.90972	28		32	.43104	.47769	2.0934	.90233	28
33	.41549	.45678	2.1892	.90960	27		33	.43130	.47805	2.0918	.90221	27
34	.41575	.45713	2.1876	.90948	26		34	.43156	.47840	2.0903	.90208	26
35	.41602	.45748	2.1859	.90936	25		35	.43182	.47876	2.0887	.90196	25
36	.41628	.45784	2.1842	.90924	24		36	.43209	.47912	2.0872	.90183	24
37	.41655	.45819	2.1825	.90911	23		37	.43235	.47948	2.0856	.90171	23
38	.41681	.45854	2.1808	.90899	22		38	.43261	.47984	2.0840	.90158	22
39	.41707	.45889	2.1792	.90887	21		39	.43287	.48019	2.0825	.90146	21
40	.41734	.45924	2.1775	.90875	20		40	.43313	.48055	2.0809	.90133	20
41	.41760	.45960	2.1758	.90863	19		41	.43340	.48091	2.0794	.90120	19
42	.41787	.45995	2.1742	.90851	18		42	.43366	.48127	2.0778	.90108	18
43	.41813	.46030	2.1725	.90839	17		43	.43392	.48163	2.0763	.90095	17
44	.41840	.46065	2.1708	.90826	16		44	.43418	.48198	2.0748	.90082	16
45	.41866	.46101	2.1692	.90814	15		45	.43445	.48234	2.0732	.90070	15
46	.41892	.46136	2.1675	.90802	14		46	.43471	.48270	2.0717	.90057	14
47	.41919	.46171	2.1659	.90790	13		47	.43497	.48306	2.0701	.90045	13
48	.41945	.46206	2.1642	.90778	12		48	.43523	.48342	2.0686	.90032	12
49	.41972	.46242	2.1625	.90766	11		49	.43549	.48378	2.0671	.90019	11
50	.41998	.46277	2.1609	.90753	10		50	.43575	.48414	2.0655	.90007	10
51	.42024	.46312	2.1592	.90741	9		51	.43602	.48450	2.0640	.89994	9
52	.42051	.46348	2.1576	.90729	8		52	.43628	.48486	2.0625	.89981	8
53	.42077	.46383	2.1560	.90717	7		53	.43654	.48521	2.0609	.89968	7
54	.42104	.46418	2.1543	.90704	6		54	.43680	.48557	2.0594	.89956	6
55	.42130	.46454	2.1527	.90692	5		55	.43706	.48593	2.0579	.89943	5
56	.42156	.46489	2.1510	.90680	4		56	.43733	.48629	2.0564	.89930	4
57	.42183	.46525	2.1494	.90668	3		57	.43759	.48665	2.0549	.89918	3
58	.42209	.46560	2.1478	.90655	2		58	.43785	.48701	2.0533	.89905	2
59	.42235	.46595	2.1461	.90643	1		59	.43811	.48737	2.0518	.89892	1
60	.42262	.46631	2.1445	.90631	0		60	.43837	.48773	2.0503	.89879	0
′	Cos	Ctn	Tan	Sin	′		′	Cos	Ctn	Tan	Sin	′

114°   **65°**   **115°**   **64°**

26°					153°	27°					152°
′	Sin	Tan	Ctn	Cos	′	′	Sin	Tan	Ctn	Cos	′
0	.43837	.48773	2.0503	.89879	60	0	.45399	.50953	1.9626	.89101	60
1	.43863	.48809	2.0488	.89867	59	1	.45425	.50989	1.9612	.89087	59
2	.43889	.48845	2.0473	.89854	58	2	.45451	.51026	1.9598	.89074	58
3	.43916	.48881	2.0458	.89841	57	3	.45477	.51063	1.9584	.89061	57
4	.43942	.48917	2.0443	.89828	56	4	.45503	.51099	1.9570	.89048	56
5	.43968	.48953	2.0428	.89816	55	5	.45529	.51136	1.9556	.89035	55
6	.43994	.48989	2.0413	.89803	54	6	.45554	.51173	1.9542	.89021	54
7	.44020	.49026	2.0398	.89790	53	7	.45580	.51209	1.9528	.89008	53
8	.44046	.49062	2.0383	.89777	52	8	.45606	.51246	1.9514	.88995	52
9	.44072	.49098	2.0368	.89764	51	9	.45632	.51283	1.9500	.88981	51
10	.44098	.49134	2.0353	.89752	50	10	.45658	.51319	1.9486	.88968	50
11	.44124	.49170	2.0338	.89739	49	11	.45684	.51356	1.9472	.88955	49
12	.44151	.49206	2.0323	.89726	48	12	.45710	.51393	1.9458	.88942	48
13	.44177	.49242	2.0308	.89713	47	13	.45736	.51430	1.9444	.88928	47
14	.44203	.49278	2.0293	.89700	46	14	.45762	.51467	1.9430	.88915	46
15	.44229	.49315	2.0278	.89687	45	15	.45787	.51503	1.9416	.88902	45
16	.44255	.49351	2.0263	.89674	44	16	.45813	.51540	1.9402	.88888	44
17	.44281	.49387	2.0248	.89662	43	17	.45839	.51577	1.9388	.88875	43
18	.44307	.49423	2.0233	.89649	42	18	.45865	.51614	1.9375	.88862	42
19	.44333	.49459	2.0219	.89636	41	19	.45891	.51651	1.9361	.88848	41
20	.44359	.49495	2.0204	.89623	40	20	.45917	.51688	1.9347	.88835	40
21	.44385	.49532	2.0189	.89610	39	21	.45942	.51724	1.9333	.88822	39
22	.44411	.49568	2.0174	.89597	38	22	.45968	.51761	1.9319	.88808	38
23	.44437	.49604	2.0160	.89584	37	23	.45994	.51798	1.9306	.88795	37
24	.44464	.49640	2.0145	.89571	36	24	.46020	.51835	1.9292	.88782	36
25	.44490	.49677	2.0130	.89558	35	25	.46046	.51872	1.9278	.88768	35
26	.44516	.49713	2.0115	.89545	34	26	.46072	.51909	1.9265	.88755	34
27	.44542	.49749	2.0101	.89532	33	27	.46097	.51946	1.9251	.88741	33
28	.44568	.49786	2.0086	.89519	32	28	.46123	.51983	1.9237	.88728	32
29	.44594	.49822	2.0072	.89506	31	29	.46149	.52020	1.9223	.88715	31
30	.44620	.49858	2.0057	.89493	30	30	.46175	.52057	1.9210	.88701	30
31	.44646	.49894	2.0042	.89480	29	31	.46201	.52094	1.9196	.88688	29
32	.44672	.49931	2.0028	.89467	28	32	.46226	.52131	1.9183	.88674	28
33	.44698	.49967	2.0013	.89454	27	33	.46252	.52168	1.9169	.88661	27
34	.44724	.50004	1.9999	.89441	26	34	.46278	.52205	1.9155	.88647	26
35	.44750	.50040	1.9984	.89428	25	35	.46304	.52242	1.9142	.88634	25
36	.44776	.50076	1.9970	.89415	24	36	.46330	.52279	1.9128	.88620	24
37	.44802	.50113	1.9955	.89402	23	37	.46355	.52316	1.9115	.88607	23
38	.44828	.50149	1.9941	.89389	22	38	.46381	.52353	1.9101	.88593	22
39	.44854	.50185	1.9926	.89376	21	39	.46407	.52390	1.9088	.88580	21
40	.44880	.50222	1.9912	.89363	20	40	.46433	.52427	1.9074	.88566	20
41	.44906	.50258	1.9897	.89350	19	41	.46458	.52464	1.9061	.88553	19
42	.44932	.50295	1.9883	.89337	18	42	.46484	.52501	1.9047	.88539	18
43	.44958	.50331	1.9868	.89324	17	43	.46510	.52538	1.9034	.88526	17
44	.44984	.50368	1.9854	.89311	16	44	.46536	.52575	1.9020	.88512	16
45	.45010	.50404	1.9840	.89298	15	45	.46561	.52613	1.9007	.88499	15
46	.45036	.50441	1.9825	.89285	14	46	.46587	.52650	1.8993	.88485	14
47	.45062	.50477	1.9811	.89272	13	47	.46613	.52687	1.8980	.88472	13
48	.45088	.50514	1.9797	.89259	12	48	.46639	.52724	1.8967	.88458	12
49	.45114	.50550	1.9782	.89245	11	49	.46664	.52761	1.8953	.88445	11
50	.45140	.50587	1.9768	.89232	10	50	.46690	.52798	1.8940	.88431	10
51	.45166	.50623	1.9754	.89219	9	51	.46716	.52836	1.8927	.88417	9
52	.45192	.50660	1.9740	.89206	8	52	.46742	.52873	1.8913	.88404	8
53	.45218	.50696	1.9725	.89193	7	53	.46767	.52910	1.8900	.88390	7
54	.45243	.50733	1.9711	.89180	6	54	.46793	.52947	1.8887	.88377	6
55	.45269	.50769	1.9697	.89167	5	55	.46819	.52985	1.8873	.88363	5
56	.45295	.50806	1.9683	.89153	4	56	.46844	.53022	1.8860	.88349	4
57	.45321	.50843	1.9669	.89140	3	57	.46870	.53059	1.8847	.88336	3
58	.45347	.50879	1.9654	.89127	2	58	.46896	.53096	1.8834	.88322	2
59	.45373	.50916	1.9640	.89114	1	59	.46921	.53134	1.8820	.88308	1
60	.45399	.50953	1.9626	.89101	0	60	.46947	.53171	1.8807	.88295	0
′	Cos	Ctn	Tan	Sin	′	′	Cos	Ctn	Tan	Sin	′

| 116° | | | | | 63° | 117° | | | | | 62° |

28°				151°		29°					150°
′	Sin	Tan	Ctn	Cos	′	′	Sin	Tan	Ctn	Cos	′
0	.46947	.53171	1.8807	.88295	60	0	.48481	.55431	1.8040	.87462	60
1	.46973	.53208	1.8794	.88281	59	1	.48506	.55469	1.8028	.87448	59
2	.46999	.53246	1.8781	.88267	58	2	.48532	.55507	1.8016	.87434	58
3	.47024	.53283	1.8768	.88254	57	3	.48557	.55545	1.8003	.87420	57
4	.47050	.53320	1.8755	.88240	56	4	.48583	.55583	1.7991	.87406	56
5	.47076	.53358	1.8741	.88226	55	5	.48608	.55621	1.7979	.87391	55
6	.47101	.53395	1.8728	.88213	54	6	.48634	.55659	1.7966	.87377	54
7	.47127	.53432	1.8715	.88199	53	7	.48659	.55697	1.7954	.87363	53
8	.47153	.53470	1.8702	.88185	52	8	.48684	.55736	1.7942	.87349	52
9	.47178	.53507	1.8689	.88172	51	9	.48710	.55774	1.7930	.87335	51
10	.47204	.53545	1.8676	.88158	50	10	.48735	.55812	1.7917	.87321	50
11	.47229	.53582	1.8663	.88144	49	11	.48761	.55850	1.7905	.87306	49
12	.47255	.53620	1.8650	.88130	48	12	.48786	.55888	1.7893	.87292	48
13	.47281	.53657	1.8637	.88117	47	13	.48811	.55926	1.7881	.87278	47
14	.47306	.53694	1.8624	.88103	46	14	.48837	.55964	1.7868	.87264	46
15	.47332	.53732	1.8611	.88089	45	15	.48862	.56003	1.7856	.87250	45
16	.47358	.53769	1.8598	.88075	44	16	.48888	.56041	1.7844	.87235	44
17	.47383	.53807	1.8585	.88062	43	17	.48913	.56079	1.7832	.87221	43
18	.47409	.53844	1.8572	.88048	42	18	.48938	.56117	1.7820	.87207	42
19	.47434	.53882	1.8559	.88034	41	19	.48964	.56156	1.7808	.87193	41
20	.47460	.53920	1.8546	.88020	40	20	.48989	.56194	1.7796	.87178	40
21	.47486	.53957	1.8533	.88006	39	21	.49014	.56232	1.7783	.87164	39
22	.47511	.53995	1.8520	.87993	38	22	.49040	.56270	1.7771	.87150	38
23	.47537	.54032	1.8507	.87979	37	23	.49065	.56309	1.7759	.87136	37
24	.47562	.54070	1.8495	.87965	36	24	.49090	.56347	1.7747	.87121	36
25	.47588	.54107	1.8482	.87951	35	25	.49116	.56385	1.7735	.87107	35
26	.47614	.54145	1.8469	.87937	34	26	.49141	.56424	1.7723	.87093	34
27	.47639	.54183	1.8456	.87923	33	27	.49166	.56462	1.7711	.87079	33
28	.47665	.54220	1.8443	.87909	32	28	.49192	.56501	1.7699	.87064	32
29	.47690	.54258	1.8430	.87896	31	29	.49217	.56539	1.7687	.87050	31
30	.47716	.54296	1.8418	.87882	30	30	.49242	.56577	1.7675	.87036	30
31	.47741	.54333	1.8405	.87868	29	31	.49268	.56616	1.7663	.87021	29
32	.47767	.54371	1.8392	.87854	28	32	.49293	.56654	1.7651	.87007	28
33	.47793	.54409	1.8379	.87840	27	33	.49318	.56693	1.7639	.86993	27
34	.47818	.54446	1.8367	.87826	26	34	.49344	.56731	1.7627	.86978	26
35	.47844	.54484	1.8354	.87812	25	35	.49369	.56769	1.7615	.86964	25
36	.47869	.54522	1.8341	.87798	24	36	.49394	.56808	1.7603	.86949	24
37	.47895	.54560	1.8329	.87784	23	37	.49419	.56846	1.7591	.86935	23
38	.47920	.54597	1.8316	.87770	22	38	.49445	.56885	1.7579	.86921	22
39	.47946	.54635	1.8303	.87756	21	39	.49470	.56923	1.7567	.86906	21
40	.47971	.54673	1.8291	.87743	20	40	.49495	.56962	1.7556	.86892	20
41	.47997	.54711	1.8278	.87729	19	41	.49521	.57000	1.7544	.86878	19
42	.48022	.54748	1.8265	.87715	18	42	.49546	.57039	1.7532	.86863	18
43	.48048	.54786	1.8253	.87701	17	43	.49571	.57078	1.7520	.86849	17
44	.48073	.54824	1.8240	.87687	16	44	.49596	.57116	1.7508	.86834	16
45	.48099	.54862	1.8228	.87673	15	45	.49622	.57155	1.7496	.86820	15
46	.48124	.54900	1.8215	.87659	14	46	.49647	.57193	1.7485	.86805	14
47	.48150	.54938	1.8202	.87645	13	47	.49672	.57232	1.7473	.86791	13
48	.48175	.54975	1.8190	.87631	12	48	.49697	.57271	1.7461	.86777	12
49	.48201	.55013	1.8177	.87617	11	49	.49723	.57309	1.7449	.86762	11
50	.48226	.55051	1.8165	.87603	10	50	.49748	.57348	1.7437	.86748	10
51	.48252	.55089	1.8152	.87589	9	51	.49773	.57386	1.7426	.86733	9
52	.48277	.55127	1.8140	.87575	8	52	.49798	.57425	1.7414	.86719	8
53	.48303	.55165	1.8127	.87561	7	53	.49824	.57464	1.7402	.86704	7
54	.48328	.55203	1.8115	.87546	6	54	.49849	.57503	1.7391	.86690	6
55	.48354	.55241	1.8103	.87532	5	55	.49874	.57541	1.7379	.86675	5
56	.48379	.55279	1.8090	.87518	4	56	.49899	.57580	1.7367	.86661	4
57	.48405	.55317	1.8078	.87504	3	57	.49924	.57619	1.7355	.86646	3
58	.48430	.55355	1.8065	.87490	2	58	.49950	.57657	1.7344	.86632	2
59	.48456	.55393	1.8053	.87476	1	59	.49975	.57696	1.7332	.86617	1
60	.48481	.55431	1.8040	.87462	0	60	.50000	.57735	1.7321	.86603	0
′	Cos	Ctn	Tan	Sin	′	′	Cos	Ctn	Tan	Sin	′

TABLE OF NATURAL TRIGONOMETRIC FUNCTIONS (Cont.)

30°   **149°**   **31°**   **148°**

′	Sin	Tan	Ctn	Cos	′	′	Sin	Tan	Ctn	Cos	′
0	.50000	.57735	1.7321	.86603	60	0	.51504	.60086	1.6643	.85717	60
1	.50025	.57774	1.7309	.86588	59	1	.51529	.60126	1.6632	.85702	59
2	.50050	.57813	1.7297	.86573	58	2	.51554	.60165	1.6621	.85687	58
3	.50076	.57851	1.7286	.86559	57	3	.51579	.60205	1.6610	.85672	57
4	.50101	.57890	1.7274	.86544	56	4	.51604	.60245	1.6599	.85657	56
5	.50126	.57929	1.7262	.86530	55	5	.51628	.60284	1.6588	.85642	55
6	.50151	.57968	1.7251	.86515	54	6	.51653	.60324	1.6577	.85627	54
7	.50176	.58007	1.7239	.86501	53	7	.51678	.60364	1.6566	.85612	53
8	.50201	.58046	1.7228	.86486	52	8	.51703	.60403	1.6555	.85597	52
9	.50227	.58085	1.7216	.86471	51	9	.51728	.60443	1.6545	.85582	51
10	.50252	.58124	1.7205	.86457	50	10	.51753	.60483	1.6534	.85567	50
11	.50277	.58162	1.7193	.86442	49	11	.51778	.60522	1.6523	.85551	49
12	.50302	.48201	1.7182	.86427	48	12	.51803	.60562	1.6512	.85536	48
13	.50327	.58240	1.7170	.86413	47	13	.51828	.60602	1.6501	.85521	47
14	.50352	.58279	1.7159	.86398	46	14	.51852	.60642	1.6490	.85506	46
15	.50377	.58318	1.7147	.86384	45	15	.51877	.60681	1.6479	.85491	45
16	.50403	.58357	1.7136	.86369	44	16	.51902	.60721	1.6469	.85476	44
17	.50428	.58396	1.7124	.86354	43	17	.51927	.60761	1.6458	.85461	43
18	.50453	.58435	1.7113	.86340	42	18	.51952	.60801	1.6447	.85446	42
19	.50478	.58474	1.7102	.86325	41	19	.51977	.60841	1.6436	.85431	41
20	.50503	.58513	1.7090	.86310	40	20	.52002	.60881	1.6426	.85416	40
21	.50528	.58552	1.7079	.86295	39	21	.52026	.60921	1.6415	.85401	39
22	.50553	.58591	1.7067	.86281	38	22	.52051	.60960	1.6404	.85385	38
23	.50578	.58631	1.7056	.86266	37	23	.52076	.61000	1.6393	.85370	37
24	.50603	.58670	1.7045	.86251	36	24	.52101	.61040	1.6383	.85355	36
25	.50628	.58709	1.7033	.86237	35	25	.52126	.61080	1.6372	.85340	35
26	.50654	.58748	1.7022	.86222	34	26	.52151	.61120	1.6361	.85325	34
27	.50679	.58787	1.7011	.86207	33	27	.52175	.61160	1.6351	.85310	33
28	.50704	.58826	1.6999	.86192	32	28	.52200	.61200	1.6340	.85294	32
29	.50729	.58865	1.6988	.86178	31	29	.52225	.61240	1.6329	.85279	31
30	.50754	.58905	1.6977	.86163	30	30	.52250	.61280	1.6319	.85264	30
31	.50779	.58944	1.6965	.86148	29	31	.52275	.61320	1.6308	.85249	29
32	.50804	.58983	1.6954	.86133	28	32	.52299	.61360	1.6297	.85234	28
33	.50829	.59022	1.6943	.86119	27	33	.52324	.61400	1.6287	.85218	27
34	.50854	.59061	1.6932	.86104	26	34	.52349	.61440	1.6276	.85203	26
35	.50879	.59101	1.6920	.86089	25	35	.52374	.61480	1.6265	.85188	25
36	.50904	.59140	1.6909	.86074	24	36	.52399	.61520	1.6255	.85173	24
37	.50929	.59179	1.6898	.86059	23	37	.52423	.61561	1.6244	.85157	23
38	.50954	.59218	1.6887	.86045	22	38	.52448	.61601	1.6234	.85142	22
39	.50979	.59258	1.6875	.86030	21	39	.52473	.61641	1.6223	.85127	21
40	.51004	.59297	1.6864	.86015	20	40	.52498	.61681	1.6212	.85112	20
41	.51029	.59336	1.6853	.86000	19	41	.52522	.61721	1.6202	.85096	19
42	.51054	.59376	1.6842	.85985	18	42	.52547	.61761	1.6191	.85081	18
43	.51079	.59415	1.6831	.85970	17	43	.52572	.61801	1.6181	.85066	17
44	.51104	.59454	1.6820	.85956	16	44	.52597	.61842	1.6170	.85051	16
45	.51129	.59494	1.6808	.85941	15	45	.52621	.61882	1.6160	.85035	15
46	.51154	.59533	1.6797	.85926	14	46	.52646	.61922	1.6149	.85020	14
47	.51179	.59573	1.6786	.85911	13	47	.52671	.61962	1.6139	.85005	13
48	.51204	.59612	1.6775	.85896	12	48	.52696	.62003	1.6128	.84989	12
49	.51229	.59651	1.6764	.85881	11	49	.52720	.62043	1.6118	.84974	11
50	.51254	.59691	1.6753	.85866	10	50	.52745	.62083	1.6107	.84959	10
51	.51279	.59730	1.6742	.85851	9	51	.52770	.62124	1.6097	.84943	9
52	.51304	.59770	1.6731	.85836	8	52	.52794	.62164	1.6087	.84928	8
53	.51329	.59809	1.6720	.85821	7	53	.52819	.62204	1.6076	.84913	7
54	.51354	.59849	1.6709	.85806	6	54	.52844	.62245	1.6066	.84897	6
55	.51379	.59888	1.6698	.85792	5	55	.52869	.62285	1.6055	.84882	5
56	.51404	.59928	1.6687	.85777	4	56	.52893	.62325	1.6045	.84866	4
57	.51429	.59967	1.6676	.85762	3	57	.52918	.62366	1.6034	.84851	3
58	.51454	.60007	1.6665	.85747	2	58	.52943	.62406	1.6024	.84836	2
59	.51479	.60046	1.6654	.85732	1	59	.52967	.62446	1.6014	.84820	1
60	.51504	.60086	1.6643	.85717	0	60	.52992	.62487	1.6003	.84805	0
′	Cos	Ctn	Tan	Sin	′	′	Cos	Ctn	Tan	Sin	′

32°					147°	33°					146°
′	Sin	Tan	Ctn	Cos	′	′	Sin	Tan	Ctn	Cos	′
0	.52992	.62487	1.6003	.84805	60	0	.54464	.64941	1.5399	.83867	60
1	.53017	.62527	1.5993	.84789	59	1	.54488	.64982	1.5389	.83851	59
2	.53041	.62568	1.5983	.84774	58	2	.54513	.65024	1.5379	.83835	58
3	.53066	.62608	1.5972	.84759	57	3	.54537	.65065	1.5369	.83819	57
4	.53091	.62649	1.5962	.84743	56	4	.54561	.65106	1.5359	.83804	56
5	.53115	.62689	1.5952	.84728	55	5	.54586	.65148	1.5350	.83788	55
6	.53140	.62730	1.5941	.84712	54	6	.54610	.65189	1.5340	.83772	54
7	.53164	.62770	1.5931	.84697	53	7	.54635	.65231	1.5330	.83756	53
8	.53189	.62811	1.5921	.84681	52	8	.54659	.65272	1.5320	.83740	52
9	.53214	.62852	1.5911	.84666	51	9	.54683	.65314	1.5311	.83724	51
10	.53238	.62892	1.5900	.84650	50	10	.54708	.65355	1.5301	.83708	50
11	.53263	.62933	1.5890	.84635	49	11	.54732	.65397	1.5291	.83692	49
12	.53288	.62973	1.5880	.84619	48	12	.54756	.65438	1.5282	.83676	48
13	.53312	.63014	1.5869	.84604	47	13	.54781	.65480	1.5272	.83660	47
14	.53337	.63055	1.5859	.84588	46	14	.54805	.65521	1.5262	.83645	46
15	.53361	.63095	1.5849	.84573	45	15	.54829	.65563	1.5253	.83629	45
16	.53386	.63136	1.5839	.84557	44	16	.54854	.65604	1.5243	.83613	44
17	.53411	.63177	1.5829	.84542	43	17	.54878	.65646	1.5233	.83597	43
18	.53435	.63217	1.5818	.84526	42	18	.54902	.65688	1.5224	.83581	42
19	.53460	.63258	1.5808	.84511	41	19	.54927	.65729	1.5214	.83565	41
20	.53484	.63299	1.5798	.84495	40	20	.54951	.65771	1.5204	.83549	40
21	.53509	.63340	1.5788	.84480	39	21	.54975	.65813	1.5195	.83533	39
22	.53534	.63380	1.5778	.84464	38	22	.54999	.65854	1.5185	.83517	38
23	.53558	.63421	1.5768	.84448	37	23	.55024	.65896	1.5175	.83501	37
24	.53583	.63462	1.5757	.84433	36	24	.55048	.65938	1.5166	.83485	36
25	.53607	.63503	1.5747	.84417	35	25	.55072	.65980	1.5156	.83469	35
26	.53632	.63544	1.5737	.84402	34	26	.55097	.66021	1.5147	.83453	34
27	.53656	.63584	1.5727	.84386	33	27	.55121	.66063	1.5137	.83437	33
28	.53681	.63625	1.5717	.84370	32	28	.55145	.66105	1.5127	.83421	32
29	.53705	.63666	1.5707	.84355	31	29	.55169	.66147	1.5118	.83405	31
30	.53730	.63707	1.5697	.84339	30	30	.55194	.66189	1.5108	.83389	30
31	.53754	.63748	1.5687	.84324	29	31	.55218	.66230	1.5099	.83373	29
32	.53779	.63789	1.5677	.84308	28	32	.55242	.66272	1.5089	.83356	28
33	.53804	.63830	1.5667	.84292	27	33	.55266	.66314	1.5080	.83340	27
34	.53828	.63871	1.5657	.84277	26	34	.55291	.66356	1.5070	.83324	26
35	.53853	.63912	1.5647	.84261	25	35	.55315	.66398	1.5061	.83308	25
36	.53877	.63953	1.5637	.84245	24	36	.55339	.66440	1.5051	.83292	24
37	.53902	.63994	1.5627	.84230	23	37	.55363	.66482	1.5042	.83276	23
38	.53926	.64035	1.5617	.84214	22	38	.55388	.66524	1.5032	.83260	22
39	.53951	.64076	1.5607	.84198	21	39	.55412	.66566	1.5023	.83244	21
40	.53975	.64117	1.5597	.84182	20	40	.55436	.66608	1.5013	.83228	20
41	.54000	.64158	1.5587	.84167	19	41	.55460	.66650	1.5004	.83212	19
42	.54024	.64199	1.5577	.84151	18	42	.55484	.66692	1.4994	.83195	18
43	.54049	.64240	1.5567	.84135	17	43	.55509	.66734	1.4985	.83179	17
44	.54073	.64281	1.5557	.84120	16	44	.55533	.66776	1.4975	.83163	16
45	.54097	.64322	1.5547	.84104	15	45	.55557	.66818	1.4966	.83147	15
46	.54122	.64363	1.5537	.84088	14	46	.55581	.66860	1.4957	.83131	14
47	.54146	.64404	1.5527	.84072	13	47	.55605	.66902	1.4947	.83115	13
48	.54171	.64446	1.5517	.84057	12	48	.55630	.66944	1.4938	.83098	12
49	.54195	.64487	1.5507	.84041	11	49	.55654	.66986	1.4928	.83082	11
50	.54220	.64528	1.5497	.84025	10	50	.55678	.67028	1.4919	.83066	10
51	.54244	.64569	1.5487	.84009	9	51	.55702	.67071	1.4910	.83050	9
52	.54269	.64610	1.5477	.83994	8	52	.55726	.67113	1.4900	.83034	8
53	.54293	.64652	1.5468	.83978	7	53	.55750	.67155	1.4891	.83017	7
54	.54317	.64693	1.5458	.83962	6	54	.55775	.67197	1.4882	.83001	6
55	.54342	.64734	1.5448	.83946	5	55	.55799	.67239	1.4872	.82985	5
56	.54366	.64775	1.5438	.83930	4	56	.55823	.67282	1.4863	.82969	4
57	.54391	.64817	1.5428	.83915	3	57	.55847	.67324	1.4854	.82953	3
58	.54415	.64858	1.5418	.83899	2	58	.55871	.67366	1.4844	.82936	2
59	.54440	.64899	1.5408	.83883	1	59	.55895	.67409	1.4835	.82920	1
60	.54464	.64941	1.5399	.83867	0	60	.55919	.67451	1.4826	.82904	0
′	Cos	Ctn	Tan	Sin	′	′	Cos	Ctn	Tan	Sin	′

| 122° | | | | | 57° | 123° | | | | | 56° |

TABLE OF NATURAL TRIGONOMETRIC FUNCTIONS (Cont.)

′	Sin	Tan	Ctn	Cos	′	′	Sin	Tan	Ctn	Cos	′
0	.55919	.67451	1.4826	.82904	60	0	.57358	.70021	1.4281	.81915	60
1	.55943	.67493	1.4816	.82887	59	1	.57381	.70064	1.4273	.81899	59
2	.55968	.67536	1.4807	.82871	58	2	.57405	.70107	1.4264	.81882	58
3	.55992	.67578	1.4798	.82855	57	3	.57429	.70151	1.4255	.81865	57
4	.56016	.67620	1.4788	.82839	56	4	.57453	.70194	1.4246	.81848	56
5	.56040	.67663	1.4779	.82822	55	5	.57477	.70238	1.4237	.81832	55
6	.56064	.67705	1.4770	.82806	54	6	.57501	.70281	1.4229	.81815	54
7	.56088	.67748	1.4761	.82790	53	7	.57524	.70325	1.4220	.81798	53
8	.56112	.67790	1.4751	.82773	52	8	.57548	.70368	1.4211	.81782	52
9	.56136	.67832	1.4742	.82757	51	9	.57572	.70412	1.4202	.81765	51
10	.56160	.67875	1.4733	.82741	50	10	.57596	.70455	1.4193	.81748	50
11	.56184	.67917	1.4724	.82724	49	11	.57619	.70499	1.4185	.81731	49
12	.56208	.67960	1.4715	.82708	48	12	.57643	.70542	1.4176	.81714	48
13	.56232	.68002	1.4705	.82692	47	13	.57667	.70586	1.4167	.81698	47
14	.56256	.68045	1.4696	.82675	46	14	.57691	.70629	1.4158	.81681	46
15	.56280	.68088	1.4687	.82659	45	15	.57715	.70673	1.4150	.81664	45
16	.56305	.68130	1.4678	.82643	44	16	.57738	.70717	1.4141	.81647	44
17	.56329	.68173	1.4669	.82626	43	17	.57762	.70760	1.4132	.81631	43
18	.56353	.68215	1.4659	.82610	42	18	.57786	.70804	1.4124	.81614	42
19	.56377	.68258	1.4650	.82593	41	19	.57810	.70848	1.4115	.81597	41
20	.56401	.68301	1.4641	.82577	40	20	.57833	.70891	1.4106	.81580	40
21	.56425	.68343	1.4632	.82561	39	21	.57857	.70935	1.4097	.81563	39
22	.56449	.68386	1.4623	.82544	38	22	.57881	.70979	1.4089	.81546	38
23	.56473	.68429	1.4614	.82528	37	23	.57904	.71023	1.4080	.81530	37
24	.56497	.68471	1.4605	.82511	36	24	.57928	.71066	1.4071	.81513	36
25	.56521	.68514	1.4596	.82495	35	25	.57952	.71110	1.4063	.81496	35
26	.56545	.68557	1.4586	.82478	34	26	.57976	.71154	1.4054	.81479	34
27	.56569	.68600	1.4577	.82462	33	27	.57999	.71198	1.4045	.81462	33
28	.56593	.68642	1.4568	.82446	32	28	.58023	.71242	1.4037	.81445	32
29	.56617	.68685	1.4559	.82429	31	29	.58047	.71285	1.4028	.81428	31
30	.56641	.68728	1.4550	.82413	30	30	.58070	.71329	1.4019	.81412	30
31	.56665	.68771	1.4541	.82396	29	31	.58094	.71373	1.4011	.81395	29
32	.56689	.68814	1.4532	.82380	28	32	.58118	.71417	1.4002	.81378	28
33	.56713	.68857	1.4523	.82363	27	33	.58141	.71461	1.3994	.81361	27
34	.56736	.68900	1.4514	.82347	26	34	.58165	.71505	1.3985	.81344	26
35	.56760	.68942	1.4505	.82330	25	35	.58189	.71549	1.3976	.81327	25
36	.56784	.68985	1.4496	.82314	24	36	.58212	.71593	1.3968	.81310	24
37	.56808	.69028	1.4487	.82297	23	37	.58236	.71637	1.3959	.81293	23
38	.56832	.69071	1.4478	.82281	22	38	.58260	.71681	1.3951	.81276	22
39	.56856	.69114	1.4469	.82264	21	39	.58283	.71725	1.3942	.81259	21
40	.56880	.69157	1.4460	.82248	20	40	.58307	.71769	1.3934	.81242	20
41	.56904	.69200	1.4451	.82231	19	41	.58330	.71813	1.3925	.81225	19
42	.56928	.69243	1.4442	.82214	18	42	.58354	.71857	1.3916	.81208	18
43	.56952	.69286	1.4433	.82198	17	43	.58378	.71901	1.3908	.81191	17
44	.56976	.69329	1.4424	.82181	16	44	.58401	.71946	1.3899	.81174	16
45	.57000	.69372	1.4415	.82165	15	45	.58425	.71990	1.3891	.81157	15
46	.57024	.69416	1.4406	.82148	14	46	.58449	.72034	1.3882	.81140	14
47	.57047	.69459	1.4397	.82132	13	47	.58472	.72078	1.3874	.81123	13
48	.57071	.69502	1.4388	.82115	12	48	.58496	.72122	1.3865	.81106	12
49	.57095	.69545	1.4379	.82098	11	49	.58519	.72167	1.3857	.81089	11
50	.57119	.69588	1.4370	.82082	10	50	.58543	.72211	1.3848	.81072	10
51	.57143	.69631	1.4361	.82065	9	51	.58567	.72255	1.3840	.81055	9
52	.57167	.69675	1.4352	.82048	8	52	.58590	.72299	1.3831	.81038	8
53	.57191	.69718	1.4344	.82032	7	53	.58614	.72344	1.3823	.81021	7
54	.57215	.69761	1.4335	.82015	6	54	.58637	.72388	1.3814	.81004	6
55	.57238	.69804	1.4326	.81999	5	55	.58661	.72432	1.3806	.80987	5
56	.47262	.69847	1.4317	.81982	4	56	.58684	.72477	1.3798	.80970	4
57	.57286	.69891	1.4308	.81965	3	57	.58708	.72521	1.3789	.80953	3
58	.57310	.69934	1.4299	.81949	2	58	.58731	.72565	1.3781	.80936	2
59	.57334	.69977	1.4290	.81932	1	59	.58755	.72610	1.3772	.80919	1
60	.57358	.70021	1.4281	.81915	0	60	.58779	.72654	1.3764	.80902	0
′	Cos	Ctn	Tan	Sin	′	′	Cos	Ctn	Tan	Sin	′

TABLE OF NATURAL TRIGONOMETRIC FUNCTIONS (Cont.)

′	Sin	Tan	Ctn	Cos	′	′	Sin	Tan	Ctn	Cos	′
0	.58779	.72654	1.3764	.80902	60	0	.60182	.75355	1.3270	.79864	60
1	.58802	.72699	1.3755	.80885	59	1	.60205	.75401	1.3262	.79846	59
2	.58826	.72743	1.3747	.80867	58	2	.60228	.75447	1.3254	.79829	58
3	.58849	.72788	1.3739	.80850	57	3	.60251	.75492	1.3246	.79811	57
4	.58873	.72832	1.3730	.80833	56	4	.60274	.75538	1.3238	.79793	56
5	.58896	.72877	1.3722	.80816	55	5	.60298	.75584	1.3230	.79776	55
6	.58920	.72921	1.3713	.80799	54	6	.60321	.75629	1.3222	.79758	54
7	.58943	.72966	1.3705	.80782	53	7	.60344	.75675	1.3214	.79741	53
8	.58967	.73010	1.3697	.80765	52	8	.60367	.75721	1.3206	.79723	52
9	.58990	.73055	1.3688	.80748	51	9	.60390	.75767	1.3198	.79706	51
10	.59014	.73100	1.3680	.80730	50	10	.60414	.75812	1.3190	.79688	50
11	.59037	.73144	1.3672	.80713	49	11	.60437	.75858	1.3182	.79671	49
12	.59061	.73189	1.3663	.80696	48	12	.60460	.75904	1.3175	.79653	48
13	.59084	.73234	1.3655	.80679	47	13	.60483	.75950	1.3167	.79635	47
14	.59108	.73278	1.3647	.80662	46	14	.60506	.75996	1.3159	.79618	46
15	.59131	.73323	1.3638	.80644	45	15	.60529	.76042	1.3151	.79600	45
16	.59154	.73368	1.3630	.80627	44	16	.60553	.76088	1.3143	.79583	44
17	.59178	.73413	1.3622	.80610	43	17	.60576	.76134	1.3135	.79565	43
18	.59201	.73457	1.3613	.80593	42	18	.60599	.76180	1.3127	.79547	42
19	.59225	.73502	1.3605	.80576	41	19	.60622	.76226	1.3119	.79530	41
20	.59248	.73547	1.3597	.80558	40	20	.60645	.76272	1.3111	.79512	40
21	.59272	.73592	1.3588	.80541	39	21	.60668	.76318	1.3103	.79494	39
22	.59295	.73637	1.3580	.80524	38	22	.60691	.76364	1.3095	.79477	38
23	.59318	.73681	1.3572	.80507	37	23	.60714	.76410	1.3087	.79459	37
24	.59342	.73726	1.3564	.80489	36	24	.60738	.76456	1.3079	.79441	36
25	.59365	.73771	1.3555	.80472	35	25	.60761	.76502	1.3072	.79424	35
26	.59389	.73816	1.3547	.80455	34	26	.60784	.76548	1.3064	.79406	34
27	.59412	.73861	1.3539	.80438	33	27	.60807	.76594	1.3056	.79388	33
28	.59436	.73906	1.3531	.80420	32	28	.60830	.76640	1.3048	.79371	32
29	.59459	.73951	1.3522	.80403	31	29	.60853	.76686	1.3040	.79353	31
30	.59482	.73996	1.3514	.80386	30	30	.60876	.76733	1.3032	.79335	30
31	.59506	.74041	1.3506	.80368	29	31	.60899	.76779	1.3024	.79318	29
32	.59529	.74086	1.3498	.80351	28	32	.60922	.76825	1.3017	.79300	28
33	.59552	.74131	1.3490	.80334	27	33	.60945	.76871	1.3009	.79282	27
34	.59576	.74176	1.3481	.80316	26	34	.60968	.76918	1.3001	.79264	26
35	.59599	.74221	1.3473	.80299	25	35	.60991	.76964	1.2993	.79247	25
36	.59622	.74267	1.3465	.80282	24	36	.61015	.77010	1.2985	.79229	24
37	.59646	.74312	1.3457	.80264	23	37	.61038	.77057	1.2977	.79211	23
38	.59669	.74357	1.3449	.80247	22	38	.61061	.77103	1.2970	.79193	22
39	.59693	.74402	1.3440	.80230	21	39	.61084	.77149	1.2962	.79176	21
40	.59716	.74447	1.3432	.80212	20	40	.61107	.77196	1.2954	.79158	20
41	.59739	.74492	1.3424	.80195	19	41	.61130	.77242	1.2946	.79140	19
42	.59763	.74538	1.3416	.80178	18	42	.61153	.77289	1.2938	.79122	18
43	.59786	.74583	1.3408	.80160	17	43	.61176	.77335	1.2931	.79105	17
44	.59809	.74628	1.3400	.80143	16	44	.61199	.77382	1.2923	.79087	16
45	.59832	.74674	1.3392	.80125	15	45	.61222	.77428	1.2915	.79069	15
46	.59856	.74719	1.3384	.80108	14	46	.61245	.77475	1.2907	.79051	14
47	.59879	.74764	1.3375	.80091	13	47	.61268	.77521	1.2900	.79033	13
48	.59902	.74810	1.3367	.80073	12	48	.61291	.77568	1.2892	.79016	12
49	.59926	.74855	1.3359	.80056	11	49	.61314	.77615	1.2884	.78998	11
50	.59949	.74900	1.3351	.80038	10	50	.61337	.77661	1.2876	.78980	10
51	.59972	.74946	1.3343	.80021	9	51	.61360	.77708	1.2869	.78962	9
52	.59995	.74991	1.3335	.80003	8	52	.61383	.77754	1.2861	.78944	8
53	.60019	.75037	1.3327	.79986	7	53	.61406	.77801	1.2853	.78926	7
54	.60042	.75082	1.3319	.79968	6	54	.61429	.77848	1.2846	.78908	6
55	.60065	.75128	1.3311	.79951	5	55	.61451	.77895	1.2838	.78891	5
56	.60089	.75173	1.3303	.79934	4	56	.61474	.77941	1.2830	.78873	4
57	.60112	.75219	1.3295	.79916	3	57	.61497	.77988	1.2822	.78855	3
58	.60135	.75264	1.3287	.79899	2	58	.61520	.78035	1.2815	.78837	2
59	.60158	.75310	1.3278	.79881	1	59	.61543	.78082	1.2807	.78819	1
60	.60182	.75355	1.3270	.79864	0	60	.61566	.78129	1.2799	.78801	0
′	Cos	Ctn	Tan	Sin	′	′	Cos	Ctn	Tan	Sin	′

TABLE OF NATURAL TRIGONOMETRIC FUNCTIONS (Cont.)

′	Sin	Tan	Ctn	Cos	′	′	Sin	Tan	Ctn	Cos	′
0	.61566	.78129	1.2799	.78801	60	0	.62932	.80978	1.2349	.77715	60
1	.61589	.78175	1.2792	.78783	59	1	.62955	.81027	1.2342	.77696	59
2	.61612	.78222	1.2784	.78765	58	2	.62977	.81075	1.2334	.77678	58
3	.61635	.78269	1.2776	.78747	57	3	.63000	.81123	1.2327	.77660	57
4	.61658	.78316	1.2769	.78729	56	4	.63022	.81171	1.2320	.77641	56
5	.61681	.78363	1.2761	.78711	55	5	.63045	.81220	1.2312	.77623	55
6	.61704	.78410	1.2753	.78694	54	6	.63068	.81268	1.2305	.77605	54
7	.61726	.78457	1.2746	.78676	53	7	.63090	.81316	1.2298	.77586	53
8	.61749	.78504	1.2738	.78658	52	8	.63113	.81364	1.2290	.77568	52
9	.61772	.78551	1.2731	.78640	51	9	.63135	.81413	1.2283	.77550	51
10	.61795	.78598	1.2723	.78622	50	10	.63158	.81461	1.2276	.77531	50
11	.61818	.78645	1.2715	.78604	49	11	.63180	.81510	1.2268	.77513	49
12	.61841	.78692	1.2708	.78586	48	12	.63203	.81558	1.2261	.77494	48
13	.61864	.78739	1.2700	.78568	47	13	.63225	.81606	1.2254	.77476	47
14	.61887	.78786	1.2693	.78550	46	14	.63248	.81655	1.2247	.77458	46
15	.61909	.78834	1.2685	.78532	45	15	.63271	.81703	1.2239	.77439	45
16	.61932	.78881	1.2677	.78514	44	16	.63293	.81752	1.2232	.77421	44
17	.61955	.78928	1.2670	.78496	43	17	.63316	.81800	1.2225	.77402	43
18	.61978	.78975	1.2662	.78478	42	18	.63338	.81849	1.2218	.77384	42
19	.62001	.79022	1.2655	.78460	41	19	.63361	.81898	1.2210	.77366	41
20	.62024	.79070	1.2647	.78442	40	20	.63383	.81946	1.2203	.77347	40
21	.62046	.79117	1.2640	.78424	39	21	.63406	.81995	1.2196	.77329	39
22	.62069	.79164	1.2632	.78405	38	22	.63428	.82044	1.2189	.77310	38
23	.62092	.79212	1.2624	.78387	37	23	.63451	.82092	1.2181	.77292	37
24	.62115	.79259	1.2617	.78369	36	24	.63473	.82141	1.2174	.77273	36
25	.62138	.79306	1.2609	.78351	35	25	.63496	.82190	1.2167	.77255	35
26	.62160	.79354	1.2602	.78333	34	26	.63518	.82238	1.2160	.77236	34
27	.62183	.79401	1.2594	.78315	33	27	.63540	.82287	1.2153	.77218	33
28	.62206	.79449	1.2587	.78297	32	28	.63563	.82336	1.2145	.77199	32
29	.62229	.79496	1.2579	.78279	31	29	.63585	.82385	1.2138	.77181	31
30	.62251	.79544	1.2572	.78261	30	30	.63608	.82434	1.2131	.77162	30
31	.62274	.79591	1.2564	.78243	29	31	.63630	.82483	1.2124	.77144	29
32	.62297	.79639	1.2557	.78225	28	32	.63653	.82531	1.2117	.77125	28
33	.62320	.79686	1.2549	.78206	27	33	.63675	.82580	1.2109	.77107	27
34	.62342	.79734	1.2542	.78188	26	34	.63698	.82629	1.2102	.77088	26
35	.62365	.79781	1.2534	.78170	25	35	.63720	.82678	1.2095	.77070	25
36	.62388	.79829	1.2527	.78152	24	36	.63742	.82727	1.2088	.77051	24
37	.62411	.79877	1.2519	.78134	23	37	.63765	.82776	1.2081	.77033	23
38	.62433	.79924	1.2512	.78116	22	38	.63787	.82825	1.2074	.77014	22
39	.62456	.79972	1.2504	.78098	21	39	.63810	.82874	1.2066	.76996	21
40	.62479	.80020	1.2497	.78079	20	40	.63832	.82923	1.2059	.76977	20
41	.62502	.80067	1.2489	.78061	19	41	.63854	.82972	1.2052	.76959	19
42	.62524	.80115	1.2482	.78043	18	42	.63877	.83022	1.2045	.76940	18
43	.62547	.80163	1.2475	.78025	17	43	.63899	.83071	1.2038	.76921	17
44	.62570	.80211	1.2467	.78007	16	44	.63922	.83120	1.2031	.76903	16
45	.62592	.80258	1.2460	.77988	15	45	.63944	.83169	1.2024	.76884	15
46	.62615	.80306	1.2452	.77970	14	46	.63966	.83218	1.2017	.76866	14
47	.62638	.80354	1.2445	.77952	13	47	.63989	.83268	1.2009	.76847	13
48	.62660	.80402	1.2437	.77934	12	48	.64011	.83317	1.2002	.76828	12
49	.62683	.80450	1.2430	.77916	11	49	.64033	.83366	1.1995	.76810	11
50	.62706	.80498	1.2423	.77897	10	50	.64056	.83415	1.1988	.76791	10
51	.62728	.80546	1.2415	.77879	9	51	.64078	.83465	1.1981	.76772	9
52	.62751	.80594	1.2408	.77861	8	52	.64100	.83514	1.1974	.76754	8
53	.62774	.80642	1.2401	.77843	7	53	.64123	.83564	1.1967	.76735	7
54	.62796	.80690	1.2393	.77824	6	54	.64145	.83613	1.1960	.76717	6
55	.62819	.80738	1.2386	.77806	5	55	.64167	.83662	1.1953	.76698	5
56	.62842	.80786	1.2378	.77788	4	56	.64190	.83712	1.1946	.76679	4
57	.62864	.80834	1.2371	.77769	3	57	.64212	.83761	1.1939	.76661	3
58	.62887	.80882	1.2364	.77751	2	58	.64234	.83811	1.1932	.76642	2
59	.62909	.80930	1.2356	.77733	1	59	.64256	.83860	1.1925	.76623	1
60	.62932	.80978	1.2349	.77715	0	60	.64279	.83910	1.1918	.76604	0
′	Cos	Ctn	Tan	Sin	′	′	Cos	Ctn	Tan	Sin	′

TABLE OF NATURAL TRIGONOMETRIC FUNCTIONS (Cont.)

40° ... **139°** | **41°** ... **138°**

′	Sin	Tan	Ctn	Cos	′	′	Sin	Tan	Ctn	Cos	′
0	.64279	.83910	1.1918	.76604	60	0	.65606	.86929	1.1504	.75471	60
1	.64301	.83960	1.1910	.76586	59	1	.65628	.86980	1.1497	.75452	59
2	.64323	.84009	1.1903	.76567	58	2	.65650	.87031	1.1490	.75433	58
3	.64346	.84059	1.1896	.76548	57	3	.65672	.87082	1.1483	.75414	57
4	.64368	.84108	1.1889	.76530	56	4	.65694	.87133	1.1477	.75395	56
5	.64390	.84158	1.1882	.76511	55	5	.65716	.87184	1.1470	.75375	55
6	.64412	.84208	1.1875	.76492	54	6	.65738	.87236	1.1463	.75356	54
7	.64435	.84258	1.1868	.76473	53	7	.65759	.87287	1.1456	.75337	53
8	.64457	.84307	1.1861	.76455	52	8	.65781	.87338	1.1450	.75318	52
9	.64479	.84357	1.1854	.76436	51	9	.65803	.87389	1.1443	.75299	51
10	.64501	.84407	1.1847	.76417	50	10	.65825	.87441	1.1436	.75280	50
11	.64524	.84457	1.1840	.76398	49	11	.65847	.87492	1.1430	.75261	49
12	.64546	.84507	1.1833	.76380	48	12	.65869	.87543	1.1423	.75241	48
13	.64568	.84556	1.1826	.76361	47	13	.65891	.87595	1.1416	.75222	47
14	.64590	.84606	1.1819	.76342	46	14	.65913	.87646	1.1410	.75203	46
15	.64612	.84656	1.1812	.76323	45	15	.65935	.87698	1.1403	.75184	45
16	.64635	.84706	1.1806	.76304	44	16	.65956	.87749	1.1396	.75165	44
17	.64657	.84756	1.1799	.76286	43	17	.65978	.87801	1.1389	.75146	43
18	.64679	.84806	1.1792	.76267	42	18	.66000	.87852	1.1383	.75126	42
19	.64701	.84856	1.1785	.76248	41	19	.66022	.87904	1.1376	.75107	41
20	.64723	.84906	1.1778	.76229	40	20	.66044	.87955	1.1369	.75088	40
21	.64746	.84956	1.1771	.76210	39	21	.66066	.88007	1.1363	.75069	39
22	.64768	.85006	1.1764	.76192	38	22	.66088	.88059	1.1356	.75050	38
23	.64790	.85057	1.1757	.76173	37	23	.66109	.88110	1.1349	.75030	37
24	.64812	.85107	1.1750	.76154	36	24	.66131	.88162	1.1343	.75011	36
25	.64834	.85157	1.1743	.76135	35	25	.66153	.88214	1.1336	.74992	35
26	.64856	.85207	1.1736	.76116	34	26	.66175	.88265	1.1329	74973	34
27	64878	.85257	1.1729	.76097	33	27	.66197	.88317	1.1323	.74953	33
28	.64901	.85308	1.1722	.76078	32	28	.66218	.88369	1.1316	.74934	32
29	.64923	.85358	1.1715	.76059	31	29	.66240	.88421	1.1310	.74915	31
30	.64945	.85408	1.1708	.76041	30	30	.66262	.88473	1.1303	.74896	30
31	.64967	.85458	1.1702	.76022	29	31	.66284	.88524	1.1296	.74876	29
32	.64989	.85509	1.1695	.76003	28	32	.66306	.88576	1.1290	.74857	28
33	.65011	.85559	1.1688	.75984	27	33	.66327	.88628	1.1283	.74838	27
34	.65033	.85609	1.1681	.75965	26	34	.66349	.88680	1.1276	.74818	26
35	.65055	.85660	1.1674	.75946	25	35	.66371	.88732	1.1270	.74799	25
36	.65077	.85710	1.1667	.75927	24	36	.66393	.88784	1.1263	.74780	24
37	.65100	.85761	1.1660	.75908	23	37	.66414	.88836	1.1257	.74760	23
38	.65122	.85811	1.1653	.75889	22	38	.66436	.88888	1.1250	.74741	22
39	.65144	.85862	1.1647	.75870	21	39	.66458	.88940	1.1243	.74722	21
40	.65166	.85912	1.1640	.75851	20	40	.66480	.88992	1.1237	.74703	20
41	.65188	.85963	1.1633	.75832	19	41	.66501	.89045	1.1230	.74683	19
42	.65210	.86014	1.1626	.75813	18	42	.66523	.89097	1.1224	.74664	18
43	.65232	.86064	1.1619	.75794	17	43	.66545	.89149	1.1217	.74644	17
44	.65254	.86115	1.1612	.75775	16	44	.66566	.89201	1.1211	.74625	16
45	.65276	.86166	1.1606	.75756	15	45	.66588	.89253	1.1204	.74606	15
46	.65298	.86216	1.1599	.75738	14	46	.66610	.89306	1.1197	.74586	14
47	.65320	.86267	1.1592	.75719	13	47	.66632	.89358	1.1191	.74567	13
48	.65342	.86318	1.1585	.75700	12	48	.66653	.89410	1.1184	.74548	12
49	.65364	.86368	1.1578	.75680	11	49	.66675	.89463	1.1178	.74528	11
50	.65386	.86419	1.1571	.75661	10	50	.66697	.89515	1.1171	.74509	10
51	.65408	.86470	1.1565	.75642	9	51	.66718	.89567	1.1165	.74489	9
52	.65430	.86521	1.1558	.75623	8	52	.66740	.89620	1.1158	.74470	8
53	.65452	.86572	1.1551	.75604	7	53	.66762	.89672	1.1152	.74451	7
54	.65474	.86623	1.1544	.75585	6	54	.66783	.89725	1.1145	.74431	6
55	.65496	.86674	1.1538	.75566	5	55	.66805	.89777	1.1139	.74412	5
56	.65518	.86725	1.1531	.75547	4	56	.66827	.89830	1.1132	.74392	4
57	.65540	.86776	1.1524	.75528	3	57	.66848	.89883	1.1126	.74373	3
58	.65562	.86827	1.1517	.75509	2	58	.66870	.89935	1.1119	.74353	2
59	.65584	.86878	1.1510	.75490	1	59	.66891	.89988	1.1113	.74334	1
60	.65606	.86929	1.1504	.75471	0	60	.66913	.90040	1.1106	.74314	0
′	Cos	Ctn	Tan	Sin	′	′	Cos	Ctn	Tan	Sin	′

130° ... **49°** | **131°** ... **48°**

42° ... 137°

′	Sin	Tan	Ctn	Cos	′
0	.66913	.90040	1.1106	.74314	60
1	.66935	.90093	1.1100	.74295	59
2	.66956	.90146	1.1093	.74276	58
3	.66978	.90199	1.1087	.74256	57
4	.66999	.90251	1.1080	.74237	56
5	.67021	.90304	1.1074	.74217	55
6	.67043	.90357	1.1067	.74198	54
7	.67064	.90410	1.1061	.74178	53
8	.67086	.90463	1.1054	.74159	52
9	.67107	.90516	1.1048	.74139	51
10	.67129	.90569	1.1041	.74120	50
11	.67151	.90621	1.1035	.74100	49
12	.67172	.90674	1.1028	.74080	48
13	.67194	.90727	1.1022	.74061	47
14	.67215	.90781	1.1016	.74041	46
15	.67237	.90834	1.1009	.74022	45
16	.67258	.90887	1.1003	.74002	44
17	.67280	.90940	1.0996	.73983	43
18	.67301	.90993	1.0990	.73963	42
19	.67323	.91046	1.0983	.73944	41
20	.67344	.91099	1.0977	.73924	40
21	.67366	.91153	1.0971	.73904	39
22	.67387	.91206	1.0964	.73885	38
23	.67409	.91259	1.0958	.73865	37
24	.67430	.91313	1.0951	.73846	36
25	.67452	.91366	1.0945	.73826	35
26	.67473	.91419	1.0939	.73806	34
27	.67495	.91473	1 0932	.73787	33
28	.67516	.91526	1.0926	.73767	32
29	.67538	.91580	1.0919	.73747	31
30	.67559	.91633	1.0913	.73728	30
31	.67580	.91687	1.0907	.73708	29
32	.67602	91740	1.0900	.73688	28
33	.67623	.91794	1.0894	.73669	27
34	.67645	.91847	1.0888	.73649	26
35	.67666	.91901	1.0881	.73629	25
36	.67688	.91955	1.0875	.73610	24
37	.67709	.92008	1.0869	.73590	23
38	.67730	.92062	1.0862	.73570	22
39	.67752	.92116	1.0856	.73551	21
40	.67773	.92170	1.0850	.73531	20
41	.67795	.92224	1.0843	.73511	19
42	.67816	.92277	1.0837	.73491	18
43	.67837	.92331	1.0831	.73472	17
44	.67859	.92385	1.0824	.73452	16
45	.67880	.92439	1.0818	.73432	15
46	.67901	.92493	1.0812	.73413	14
47	.67923	.92547	1.0805	.73393	13
48	.67944	.92601	1.0799	.73373	12
49	.67965	.92655	1.0793	.73353	11
50	.67987	.92709	1.0786	.73333	10
51	.68008	.92763	1.0780	.73314	9
52	.68029	.92817	1.0774	.73294	8
53	.68051	.92872	1.0768	.73274	7
54	.68072	.92926	1.0761	.73254	6
55	.68093	.92980	1.0755	.73234	5
56	.68115	.93034	1.0749	.73215	4
57	.68136	.93088	1.0742	.73195	3
58	.68157	.93143	1.0736	.73175	2
59	.68179	.93197	1.0730	.73155	1
60	.68200	.93252	1.0724	.73135	0
′	Cos	Ctn	Tan	Sin	′

132° ... 47°

43° ... 136°

′	Sin	Tan	Ctn	Cos	′
0	.68200	.93252	1.0724	.73135	60
1	.68221	.93306	1.0717	.73116	59
2	.68242	.93360	1.0711	.73096	58
3	.68264	.93415	1.0705	.73076	57
4	.68285	.93469	1.0699	.73056	56
5	.68306	.93524	1.0692	.73036	55
6	.68327	.93578	1.0686	.73016	54
7	.68349	.93633	1.0680	.72996	53
8	.68370	.93688	1.0674	.72976	52
9	.68391	.93742	1.0668	.72957	51
10	.68412	.93797	1.0661	.72937	50
11	.68434	.93852	1.0655	.72917	49
12	.68455	.93906	1.0649	.72897	48
13	.68476	.93961	1.0643	.72877	47
14	.68497	.94016	1.0637	.72857	46
15	.68518	.94071	1.0630	.72837	45
16	.68539	.94125	1.0624	.72817	44
17	.68561	.94180	1.0618	.72797	43
18	.68582	.94235	1.0612	.72777	42
19	.68603	.94290	1.0606	.72757	41
20	.68624	.94345	1.0599	.72737	40
21	.68645	.94400	1.0593	.72717	39
22	.68666	.94455	1.0587	.72697	38
23	.68688	.94510	1.0581	.72677	37
24	.68709	.94565	1.0575	.72657	36
25	.68730	.94620	1.0569	.72637	35
26	.68751	.94676	1.0562	.72617	34
27	.68772	.94731	1.0556	.72597	33
28	.68793	.94786	1.0550	.72577	32
29	.68814	.94841	1.0544	.72557	31
30	.68835	.94896	1.0538	.72537	30
31	.68857	.94952	1.0532	.72517	29
32	.68878	.95007	1.0526	.72497	28
33	.68899	.95062	1.0519	.72477	27
34	.68920	.95118	1.0513	.72457	26
35	.68941	.95173	1.0507	.72437	25
36	.68962	.95229	1.0501	.72417	24
37	.68983	.95284	1.0495	.72397	23
38	.69004	.95340	1.0489	.72377	22
39	.69025	.95395	1.0483	.72357	21
40	.69046	.95451	1.0477	.72337	20
41	.69067	.95506	1.0470	.72317	19
42	.69088	.95562	1.0464	.72297	18
43	.69109	.95618	1.0458	.72277	17
44	.69130	.95673	1.0452	.72257	16
45	.69151	.95729	1.0446	.72236	15
46	.69172	.95785	1.0440	.72216	14
47	.69193	.95841	1.0434	.72196	13
48	.69214	.95897	1.0428	.72176	12
49	.69235	.95952	1.0422	.72156	11
50	.69256	.96008	1.0416	.72136	10
51	.69277	.96064	1.0410	.72116	9
52	.69298	.96120	1.0404	.72095	8
53	.69319	.96176	1.0398	.72075	7
54	.69340	.96232	1.0392	.72055	6
55	.69361	.96288	1.0385	.72035	5
56	.69382	.96344	1.0379	.72015	4
57	.69403	.96400	1.0373	.71995	3
58	.69424	.96457	1.0367	.71974	2
59	.69445	.96513	1.0361	.71954	1
60	.69466	.96569	1.0355	.71934	0
′	Cos	Ctn	Tan	Sin	′

133° ... 46°

TABLE OF NATURAL TRIGONOMETRIC FUNCTIONS

′	Sin	Tan	Ctn	Cos	′
0	.69466	.96569	1.0355	.71934	60
1	.69487	.96625	1.0349	.71914	59
2	.69508	.96681	1.0343	.71894	58
3	.69529	.96738	1.0337	.71873	57
4	.69549	.96794	1.0331	.71853	56
5	.69570	.96850	1.0325	.71833	55
6	.69591	.96907	1.0319	.71813	54
7	.69612	.96963	1.0313	.71792	53
8	.69633	.97020	1.0307	.71772	52
9	.69654	.97076	1.0301	.71752	51
10	.69675	.97133	1.0295	.71732	50
11	.69696	.97189	1.0289	.71711	49
12	.69717	.97246	1.0283	.71691	48
13	.69737	.97302	1.0277	.71671	47
14	.69758	.97359	1.0271	.71650	46
15	.69779	.97416	1.0265	.71630	45
16	.69800	.97472	1.0259	.71610	44
17	.69821	.97529	1.0253	.71590	43
18	.69842	.97586	1.0247	.71569	42
19	.69862	.97643	1.0241	.71549	41
20	.69883	.97700	1.0235	.71529	40
21	.69904	.97756	1.0230	.71508	39
22	.69925	.97813	1.0224	.71488	38
23	.69946	.97870	1.0218	.71468	37
24	.69966	.97927	1.0212	.71447	36
25	.69987	.97984	1.0206	.71427	35
26	.70008	.98041	1.0200	.71407	34
27	.70029	.98098	1.0194	.71386	33
28	.70049	.98155	1.0188	.71366	32
29	.70070	.98213	1.0182	.71345	31
30	.70091	.98270	1.0176	.71325	30
31	.70112	.98327	1.0170	.71305	29
32	.70132	.98384	1.0164	.71284	28
33	.70153	.98441	1.0158	.71264	27
34	.70174	.98499	1.0152	.71243	26
35	.70195	.98556	1.0147	.71223	25
36	.70215	.98613	1.0141	.71203	24
37	.70236	.98671	1.0135	.71182	23
38	.70257	.98728	1.0129	.71162	22
39	.70277	.98786	1.0123	.71141	21
40	.70298	.98843	1.0117	.71121	20
41	.70319	.98901	1.0111	.71100	19
42	.70339	.98958	1.0105	.71080	18
43	.70360	.99016	1.0099	.71059	17
44	.70381	.99073	1.0094	.71039	16
45	.70401	.99131	1.0088	.71019	15
46	.70422	.99189	1.0082	.70998	14
47	.70443	.99247	1.0076	.70978	13
48	.70463	.99304	1.0070	.70957	12
49	.70484	.99362	1.0064	.70937	11
50	.70505	.99420	1.0058	.70916	10
51	.70525	.99478	1.0052	.70896	9
52	.70546	.99536	1.0047	.70875	8
53	.70567	.99594	1.0041	.70855	7
54	.70587	.99652	1.0035	.70834	6
55	.70608	.99710	1.0029	.70813	5
56	.70628	.99768	1.0023	.70793	4
57	.70649	.99826	1.0017	.70772	3
58	.70670	.99884	1.0012	.70752	2
59	.70690	.99942	1.0006	.70731	1
60	.70711	1.0000	1.0000	.70711	0
′	Cos	Ctn	Tan	Sin	′

index